T0122193

Lecture Notes on Data Engineering and Communications Technologies

Volume 150

Series Editor

Fatos Xhafa, Technical University of Catalonia, Barcelona, Spain

The aim of the book series is to present cutting edge engineering approaches to data technologies and communications. It will publish latest advances on the engineering task of building and deploying distributed, scalable and reliable data infrastructures and communication systems.

The series will have a prominent applied focus on data technologies and communications with aim to promote the bridging from fundamental research on data science and networking to data engineering and communications that lead to industry products, business knowledge and standardisation.

Indexed by SCOPUS, INSPEC, EI Compendex.

All books published in the series are submitted for consideration in Web of Science.

More information about this series at http://www.springer.com/bookseries/15362

Loon Ching Tang · Hongzhi Wang
Editors

Big Data Management and Analysis for Cyber Physical Systems

Selected Papers of BDET 2022

Editors
Loon Ching Tang
Department of Industrial Systems
Engineering and Management
National University of Singapore
Singapore, Singapore

Hongzhi Wang
Computer Science and Technology
Harbin Institute of Technology
Harbin, China

ISSN 2367-4512 ISSN 2367-4520 (electronic)
Lecture Notes on Data Engineering and Communications Technologies
ISBN 978-3-031-17547-3 ISBN 978-3-031-17548-0 (eBook)
https://doi.org/10.1007/978-3-031-17548-0

© The Editor(s) (if applicable) and The Author(s), under exclusive license
to Springer Nature Switzerland AG 2023
This work is subject to copyright. All rights are solely and exclusively licensed by the Publisher, whether the whole or part of the material is concerned, specifically the rights of translation, reprinting, reuse of illustrations, recitation, broadcasting, reproduction on microfilms or in any other physical way, and transmission or information storage and retrieval, electronic adaptation, computer software, or by similar or dissimilar methodology now known or hereafter developed.
The use of general descriptive names, registered names, trademarks, service marks, etc. in this publication does not imply, even in the absence of a specific statement, that such names are exempt from the relevant protective laws and regulations and therefore free for general use.
The publisher, the authors, and the editors are safe to assume that the advice and information in this book are believed to be true and accurate at the date of publication. Neither the publisher nor the authors or the editors give a warranty, expressed or implied, with respect to the material contained herein or for any errors or omissions that may have been made. The publisher remains neutral with regard to jurisdictional claims in published maps and institutional affiliations.

This Springer imprint is published by the registered company Springer Nature Switzerland AG
The registered company address is: Gewerbestrasse 11, 6330 Cham, Switzerland

Preface

In the past four years, BDET has achieved fruitful results in promoting exchanges of research ideas in big data technologies despite the challenges brought about by the coronavirus pandemic. Our mission of facilitating communications and collaborations between academia and industry has begun to take shape. As a representative of the conference organization, we are very pleased that this year's conference can be held as scheduled, and I sincerely thank all colleagues, staff and participants for their support and contributions to the BDET conference.

This volume contains the collection of full papers of the 2022 4th International Conference on Big Data Engineering and Technology (BDET 2022), which was held virtually during April 22–24, 2022. We appreciate the sharing from the invited speakers who had given interesting presentation on topics that are significant to the advances in this field. We would also like to express our gratitude to all the speakers and delegates for their insightful talks and discussions at the conference.

Notably, parallel sessions on a wide range of topics and issues, including contributed papers, all with a central focus on big data engineering and technology, such as data model and algorithm, computer and information science, etc., have been well received. These sessions cover exciting new topics which are of interest to both researchers and practitioners working on issues related to big data technologies. We look forward to seeing more scholars and having more active discussions at future conferences which are likely to be held offline as the pandemic subsided.

On behalf of the Organizing Committee, we would like to take this opportunity to thank all the authors who have submitted quality papers to the conference, as well as the advisory committee members, the program committee members, the technical committee members and external reviewers for their hard work. They have made it possible for us to put together a high-quality conference this year. Special thanks go to all participants who have contributed to the success of BDET 2022 but whose names may not have been mentioned.

Sincerely yours,

Tang Loon Ching
Conference Chair of BDET 2022

Organization

Committees

International Advisory Committees

Jay Guo	University of Technology Sydney, Australia
Sharad Mehrotra	University of California, Irvine, USA

Conference Chairs

Tang Loon Ching	National University of Singapore, Singapore
Liyanage C. De Silva	Universiti Brunei Darussalam, Brunei Darussalam

Program Chairs

Hongmei Chen	Southwest Jiaotong University, China
Hongzhi Wang	Harbin Institute of Technology, China
Fugee Tsung	Hong Kong University of Science and Technology (HKUST), Hong Kong, China

Publication Chair

Zeljko Zilic	McGill University, Canada

Technical Committee

Alaa Sheta	Southern Connecticut State University, USA
Antonio Munoz	University of Malaga, Spain
Ah-Lian Kor	Leeds Beckett University, UK
Ben Chen	Chinese University of Hong Kong, Hong Kong, China

Ming Hour Yang	Chung Yuan Christian University, Taiwan
Abdel-Badeeh M. Salem	Ain Shams University, Cairo, Egypt
Alexander Ryjov	Lomonosov Moscow State University and Russian Presidential Academy of National Economy and Public Administration, Russia
Kolla Bhanu Prakash	Koneru Lakshmaiah Education Foundation, India
Nageswara Rao Moparthi	K L University, India
Abbas Fadhil Aljuboori	University of Information Technology and Communications, Iraq
Leon Abdillah	Bina Darma University, Australia
Michal Krátký	Technical University of Ostrava, Czech Republic
Ming-Yen Lin	Feng Chia University, Taiwan
Hsiao-Wei Hu	Soochow University, Taiwan
Sheak Rashed Haider Noori	Daffodil International University, Bangladesh
Akmaral Kuatbayeva	International IT University, Kazakhstan
Asst. Ka-Chun Wong	City University of Hong Kong, Hong Kong
Asst. Liangjing Yang	Zhejiang University, China
Asst. Jyotir Moy Chatterjee	IT department at Lord Buddha Education Foundation, Nepal
Asst. Phongsak Phakamach	Rajamangala University of Technology, Thailand
Biba Josef	Universität der Bundeswehr München, Germany
Alexander Gelbukh	CIC, Instituto Politécnico Nacional, Mexico
Anthony Simonet	iExec Company, France

Contents

Modern Information Technology and Application

Data Model and Computation

Passenger Flow Control in Subway Station with Card-Swiping Data

Qian Ni[1](✉) and Yida Guo[2]

[1] Chongqing Jiaotong University, Chongqing 400074, China
nq13883974553@163.com
[2] YGSOFT Company, Zhuhai 519085, China

Abstract. Congestion is a great concern for urban rail transit, because it has a great impact on commuting efficiency and passenger safety. In order to mitigate congestion, this paper proposes a passenger flow control framework. Its function is to predict passenger number with given data and algorithms, identify possible overcrowding in advance, implement control strategies, and therefore reduce congestion. A subway station model based on the AnyLogic software is built to verify the control framework and to simulate the movement of passengers and trains. Simulation results demonstrate that the control framework has a good effect on congestion alleviation at a station level.

Keywords: Rail transit · Passenger flow control · Congestion

1 Introduction

The urban rail transit plays an indispensable role in public transportation in big cities. The passenger density reached 4–5 persons/m^2, a safety standard value, in crowded areas in many metro stations during peak hours [1]. Therefore, in order to relieve congestion, it is necessary for researchers to come up with control strategies in a station.

There is some literature on controlling passenger flow at a station level. The optimal solution for passenger flow control was to transport most passengers, a station capacity problem, and also to ensure acceptable security and comfort level [2]. Moreover, the security and comfort level fell into the domain of level of service (LOS) [3]. A passenger flow control model was proposed based on a new concept of station service capacity with LOS in various passenger demand scenarios [1]. A feedback control model was realized by computing the optimal passenger inflow of station facilities and adjusting the walking velocities to improve capacity-oriented management [4]. A passenger flow control method was presented to mitigate train delays by adjusting the opening speed of automatic fare gates [5].

However, these control models fail to consider the effect of passenger distribution in a station. Therefore, it is different to deal with the congestion in some areas in a station. Accordingly, control points were introduced in order to keep the number of passengers in key areas, like station halls, under a safety limit [6]. Nevertheless, this

© The Author(s), under exclusive license to Springer Nature Switzerland AG 2023
L. C. Tang and H. Wang (Eds.): BDET 2022, LNDECT 150, pp. 3–9, 2023.
https://doi.org/10.1007/978-3-031-17548-0_1

method has a limitation of holding too many passengers between service facilities with limited capacities. Besides, the selection of control points is subjective.

In sum, it is still difficult to efficiently control passenger distribution and reduce congestion at a station level. Hence, a control framework is proposed in this research to improve passenger distribution in a metro station. The card-swiping data from Chongqing Rail Transit Line 3 in 2018 is applied in the framework and used to forestall congestion. The control framework is verified using the AnyLogic software. The main contributions of the paper are as follows: The effect of passenger distribution in a station on congestion is considered. A passenger flow control framework with two control strategies are proposed to optimize the distribution of passengers and relieve congestion in a metro station.

The rest of this paper is organized as follows: Section "Methodology" introduces a passenger flow control framework. Section "Case Study" presents a case study on the Jiazhoulu Station of Chongqing Rail Transit Line 3. Section "Conclusions" summarizes this paper and suggests future research directions.

2 Methodology

The passenger flow control framework at a station level is shown in Fig. 1.

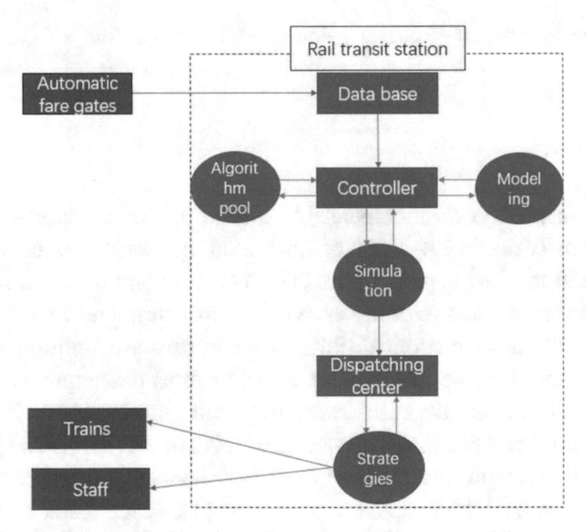

Fig. 1. Schematic diagram of passenger flow control framework

Firstly, a database will be generated from the data collected by automatic fare gates in a rail transit station. Furthermore, prediction algorithms in the algorithm pool of the controller (as is shown in Fig. 1) are used to predict the number of passengers in the following 10 min. There are plenty of prediction algorithms, including BPNN (back-propagation neural network), LSTM (Long Short-Term Memory), Transformer, etc. In order to find the algorithm that predicts passenger numbers most accurately, the database first will be divided into a training set and a test set. Then, the training set will be used to determine parameters in algorithms. Thus, the coefficient of determination will be utilized to assess the performance of every algorithm with given parameters, when the test set is inputted into algorithms. Accordingly, a best suited algorithm will be selected for every station according to the coefficient of determination. Based on the chosen algorithm, the predicted number of passengers of next 10 min will be obtained. Then, the number will be inputted into the station model on the AnyLogic. With the passenger number and the station model, a simulation result of next 10 min will be procured. Therefore, the density maps of the result could indicate the main areas of congestion. As a result, the controller alerts and sends information to the dispatching center. There are two strategies in the center to forestall incoming congestion. When there is congestion in some areas between service facilities, staff members can distribute passengers in the possible areas by guiding passenger flow so as to avoid potential risks. Another strategy is to reduce the train headway, when congestion occurs at the level of train departure. In this way, waiting passengers can get on trains quicker, which in turn relieve congestion in waiting areas.

3 Case Study

3.1 An Introduction of the Model Demo Layout

Based on the AnyLogic, a rail transit station model demo is constructed, as is shown in Figs. 2 and 3. The demo is a two-level underground station with four entrances and exits, namely A, B, C and D. On B1 floor, A and B share 6 ticket vending machines, one security check facility, five entry gates and one escalator to B2 floor. C and D have the same condition. Besides, there are five exit gates on the left side for A and B, and five exit gates on the right side for C and D. On B2 floor, as is demonstrated in Fig. 3, there are two opposite directions for taking trains.

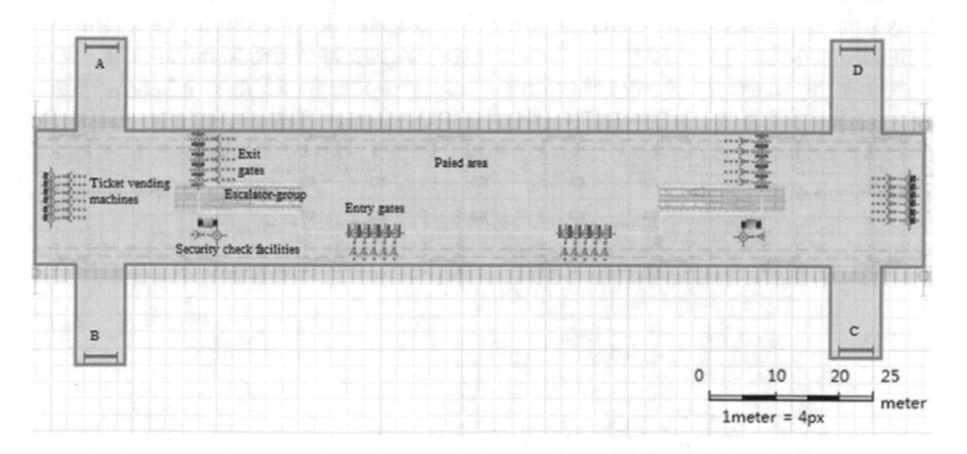

Fig. 2. A plan view of B1 floor

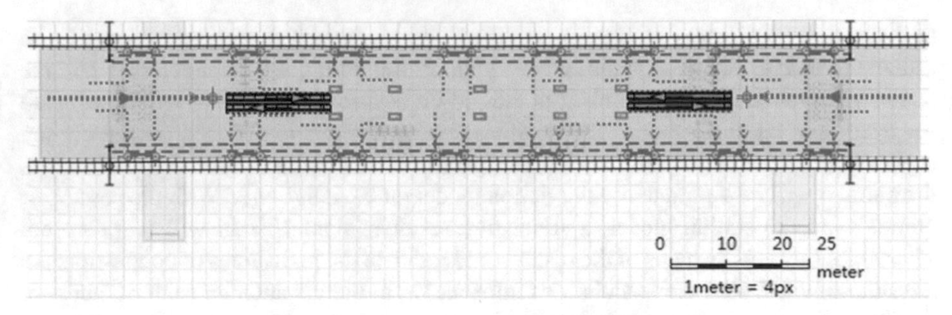

Fig. 3. A plan view of B2 floor

3.2 Data Description

The model demo is applied to the card-swiping data collected by automatic fare gates (entry gates and exit gates) from the Jiazhoulu Station of Chongqing Rail Transit Line 3 in China. The data are from the start of June to the end of December of the year 2018. Besides, the data are aggregated in 10-min intervals in order to count the number of passengers every 10 min.

3.3 Simulation and Evaluation of the Demo

The logical flowchart of inbound passengers is shown in Fig. 4. The logical flowchart of outbound passengers is demonstrated in Fig. 5. Furthermore, the logical flowchart of two trains is illustrated in Fig. 6. A specific period is simulated, i.e. from 18:10:00 to 18:20:00 on December 12, 2018. According to the data, the arrival rate of inbound passengers in the simulation is set as 1000 passengers per 10-min.

Moreover, 30% of the inbound passengers would buy tickets on ticket vending machines whose service delay time is uniformly distributed from 10 to 15 s. The service delay time for each security check facility is 2 to 3 s. Besides, the service delay time for each automatic fare gate is 2 to 3 s. What's more, the escalator-group (an escalator from B1 to B2 and an escalator from B2 to B1) runs as a speed of 0.5 m/s. Whenever trains arrive in the station and delay for 30 s, there is 200 outbound passengers who choose the nearest escalator to B1 floor. The probability for each outbound passenger of going to

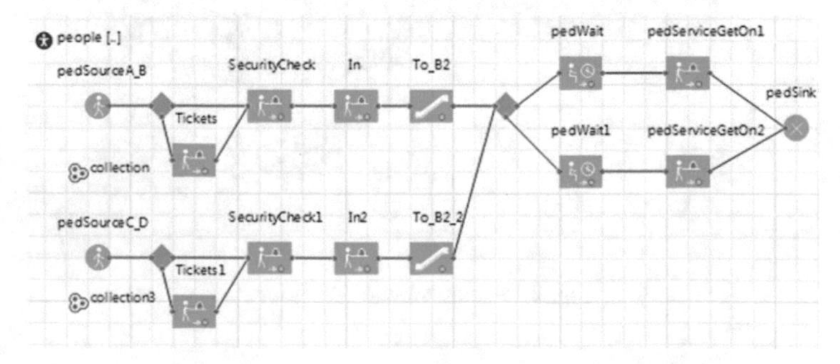

Fig. 4. Behavior modeling of inbound passengers

A, B, C or D is same. Simultaneously, the inbound passengers in the waiting areas will have 30 s to get on the trains at given doors.

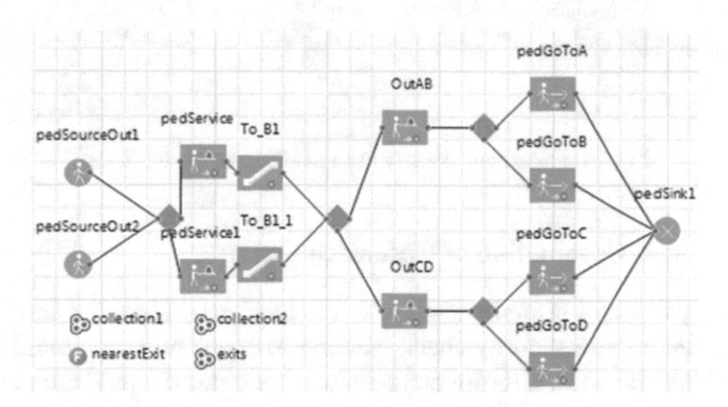

Fig. 5. Behavior modeling of outbound passengers

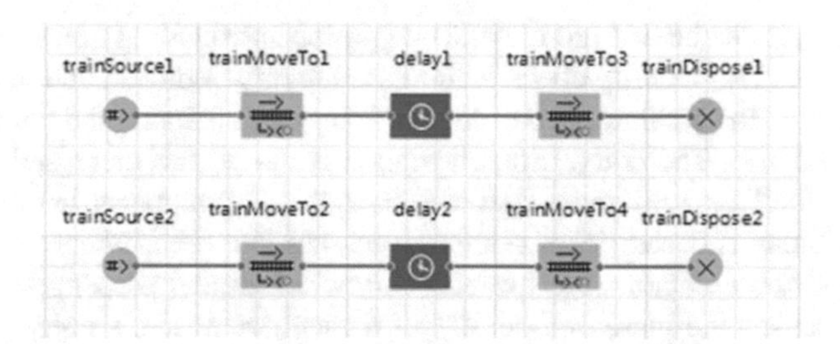

Fig. 6. Behavior modeling of of two trains

The density maps of the next 10 min are utilized to identify the main areas of congestion, as is shown in Figs. 7 and 8. Firstly, It can be seen from Fig. 7 that the main congestion on B1 floor are concentrated around areas between ticket vending machines and security check facilities. As for B2 floor in Fig. 8, the high-density points are scattered along nearly all pathways of outbound passengers, forming high-density areas near escalator-groups and waiting areas.

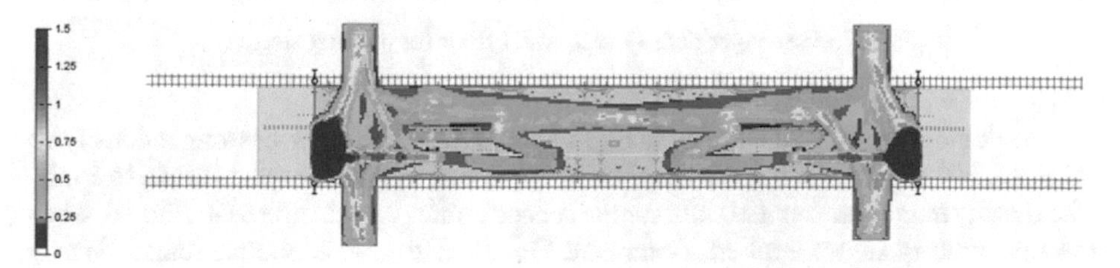

Fig. 7. Passenger density map of B1 floor of the demo

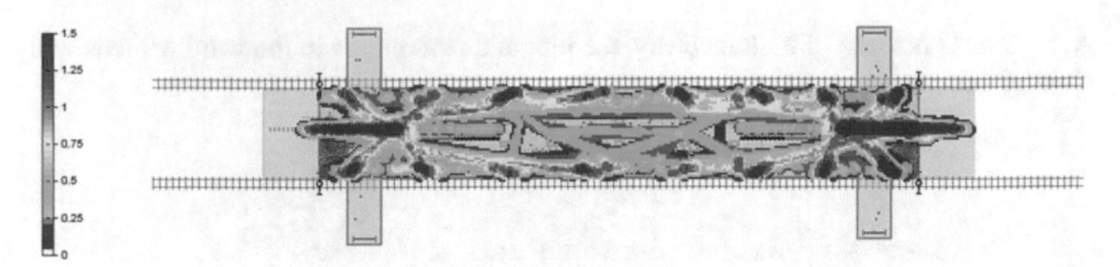

Fig. 8. Passenger density map of B2 floor of the demo

3.4 Simulation and Evaluation of Optimized Demo

The first strategy in Fig. 1 is applied in B1 floor of the demo. Thus, the first batch of staff members is arranged near security check facilities to create two new lines for passenger security checking. Besides, after inbound passengers enter the paid area through entry gates, another batch of staff will distribute passengers into two lines to take the escalator to B2 floor. The density map of the next 10 min for the first strategy is illustrated in Fig. 9.

From Fig. 9, it can be derived that the congestion surrounding areas between ticket vending machines and security check facilities are greatly reduced, and that the density of passengers near the escalator-group and the entry gates is increased compared to that in the original demo. The congestion alleviation is due to the increased efficiency of security check facilities with more staff members. In addition, because of the release of passengers in areas between ticket vending machines and security check facilities, the entry gates and the escalators are loaded with more inbound passengers. As a result, this scenario causes the increased density of passengers near the entry gates and the escalators, when the capacity of these service facilities is unchanged. Considering the whole picture, the first strategy significantly relieves congestion in the B1 floor.

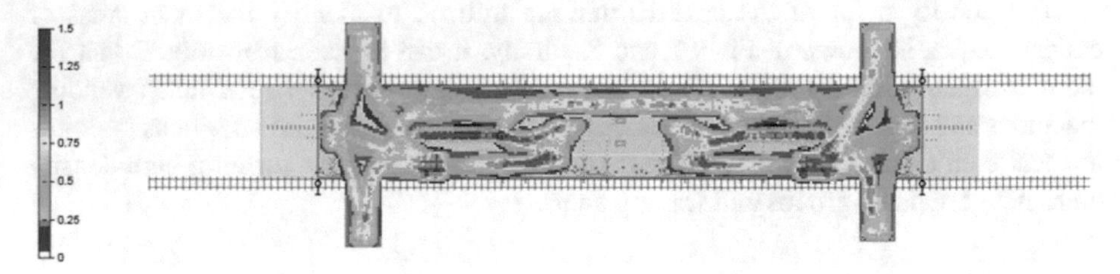

Fig. 9. Passenger density map of B1 floor for the first strategy

In addition, the second strategy in Fig. 1 is used to decrease the passenger density on B2 floor. In this case, the interarrival time between two trains is shorten from 6 to 3 min. The density map of the next 10 min for the second strategy is illustrated in Fig. 10, when the first strategy is not applied. Compared Fig. 10 with Fig. 8, the passenger density on B2 floor is decreased evidently, which proves that the second strategy has a good outcome. However, there is still long lines near the escalators due to the limited capacity of escalators.

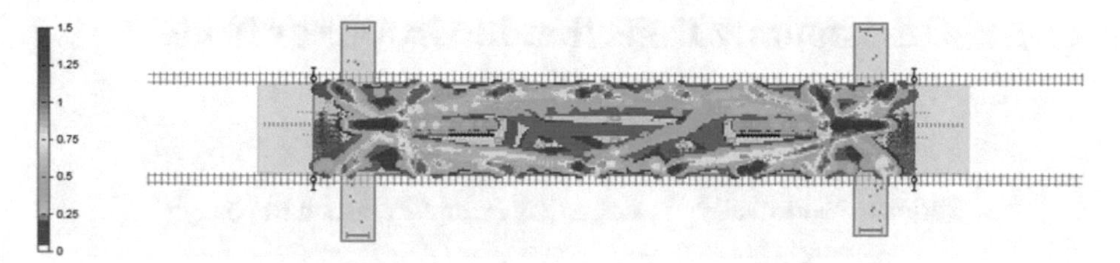

Fig. 10. Passenger density map of B2 floor for the second strategy

4 Conclusions

A new passenger flow control framework is proposed in this research. Unlike most previous models, the passenger flow control framework considers the effect of passenger distribution on congestion and has two strategies to guide passenger flow in high-density areas in a metro station. Based on the framework, congestion is alleviated at a station level. An illustrate example with a rail station demo in AnyLogic and the card-swiping data from the Jiazhoulu Station is given to demonstrate the performance of the framework. By comparing the changes in passenger density in density maps before and after implementing the strategies, the effectiveness of the framework is verified. For future work, the amount of time that each passenger spends in a station need to be considered in order to improve the travel efficiency for every passenger. Besides, there might be other strategies that can be combined with current strategies to better relieve congestion.

References

1. Xu, X., Liu, J., Li, H., Jiang, M.: Capacity-oriented passenger flow control under uncertain demand: algorithm development and real-world case study. Transp. Res. Part E Logistics Transp. Rev. **87**, 130–148 (2016). https://doi.org/10.1016/j.tre.2016.01.004
2. Yang, H., Bell, M.G.H., Meng, Q.: Modeling the capacity and level of service of urban transportation networks. Transp. Res. Part B Methodol. **34**, 255–275 (2000). https://doi.org/10.1016/S0191-2615(99)00024-7
3. Hsu, C.-I., Chao, C.-C.: Space allocation for commercial activities at international passenger terminals. Transp. Res. Part E Logistics Transp. Rev. **41**, 29–51 (2005). https://doi.org/10.1016/j.tre.2004.01.001
4. Zhang, Z., Jia, L., Qin, Y., Yun, T.: Optimization-based feedback control of passenger flow in subway stations for improving level of service. Transp. Lett. **11**, 413–424 (2019). https://doi.org/10.1080/19427867.2017.1374501
5. Yoo, S., Kim, H., Kim, W., Kim, N., Lee, J.: Controlling passenger flow to mitigate the effects of platform overcrowding on train dwell time. J. Intell. Transp. Syst. **26**(3), 366–381 (2020)
6. Wang, Y., Qin, Y., Guo, J., Jia, L., Wei, Y., Pi, Y.: Multiposition joint control in transfer station considering the nonlinear characteristics of passenger flow. J. Transp. Eng. Part A Syst. **147**, 04021068 (2021)https://doi.org/10.1061/JTEPBS.0000564

On Anomaly Detection in Graphs as Node Classification

Farimah Poursafaei[1,3(✉)], Zeljko Zilic[1], and Reihaneh Rabbany[2,3]

[1] ECE, McGill University, Montreal, Canada
zeljko.zilic@mcgill.ca
[2] CSC, McGill University, Montreal, Canada
rrabba@cs.mcgill.ca
[3] Mila, Quebec Artificial Intelligence Institute, Montreal, Canada
farimah.poursafaei@mila.quebec

Abstract. Graphs have been exercised as appealing candidates for modeling relational datasets in different domains such as cryptocurrency transaction networks, social networks, rating platforms, and many more. Recently, different powerful methods have emerged to analyze datasets where the complex underlying data connectivity can be modeled by graphs. These methods have demonstrated promising performance on graph common tasks including node classification. In this paper, we explore the impact of graph-based techniques for detecting anomalous entities on real-world networks. We focus on modeling the problem of detecting anomalous entities on a network as a node classification task, and inspect the role of different approaches together with the evaluation setup and metrics to provide several useful recommendations for practical applications. We investigate different ways of handling the imbalance issue of the datasets which is a common problem when dealing with datasets containing anomalies, and demonstrate how a method that is agnostic to the dataset imbalance may show misleading performance. Through extensive experiments on six real-world datasets in *balanced* and *unbalanced* setting for a node classification task, we provide several recommendations that can shed more lights on challenges of selecting the appropriate methods, settings, and performance metrics that better align with the intrinsic attributes of a specific dataset and task.

Keywords: Graph analysis · Anomaly detection · Node classification

1 Introduction

In the present epoch of big data, many real-world phenomena can be explored and represented through the unifying abstractions offered by graphs. In many diverse and complex data exploration and management ecosystems, big graphs processing has emerged as a principal computing framework with applications in many domains including security, social networks, finance, and many more [15]. Considering that in many use cases, the big data consists of relations, as well as vectors of features, a vital challenge is to leverage information embedded in interconnected data that is modelled by graphs.

© The Author(s), under exclusive license to Springer Nature Switzerland AG 2023
L. C. Tang and H. Wang (Eds.): BDET 2022, LNDECT 150, pp. 10–20, 2023.
https://doi.org/10.1007/978-3-031-17548-0_2

One main category of tasks in network analysis is node classification [5,6]. The task of node classification involves classifying each node of a network into one of predefined sets of classes [2]. When modeling node classification as a supervised machine learning task, the node representations can be employed by any off-the-shelf machine learning classifier to predict the classes of the nodes [5].

Node classification can be also employed for detecting anomalous entities on networks. For instance, detecting malicious users in financial networks [14], detecting fraudulent users in rating platforms that give fake rating for monetary outcomes [8], or spammer in social or financial networks [14] have all been modeled as machine learning tasks where the ultimate objective is to efficiently predict the node classes using node representations as feature vectors. In fact, many of the datasets that are extensively being used in node classification tasks contains one class of nodes that is associated with anomalous activities [8,16,17]. Hence, the node classification task on these networks are indeed a supervised anomaly detection task.

Essentially, anomalous instances refer to those that considerably deviate from seemingly normal observations [10]. Although most of classification methods address the problem in a relatively balanced setting, real-world scenarios often present datasets where some classes have considerably fewer number of instances. Training the classifiers unaware of the of the intrinsic imbalance of the datasets may results in under-representation of instances from the minority class and consequently, sub-optimal performance of the classification task [18]. In this work, we explore the application of supervised methods for classification of unbalanced datasets and demonstrate the importance of considering the intrinsic imbalance of the instances.

Imbalance problem is one of the greatest issues in data mining which relates to the case that one of the classes have considerably less number of the instances compared to others [9]. The classification methods, if overlooking the imbalance issue, mostly focus on the samples from the majority class and aim to optimize classification accuracy, while ignoring or misclassifying minority samples [9]. This becomes a vital drawback when applying classification for anomaly detection. For one thing, the datasets including anomalies are extremely imbalance which results in poor performance of the methods. For another thing, although the minority samples are rare , there are extremely important to be detected and predicting false negatives could be very costly, for instance, credit card fraud detection or detecting faults in safety critical systems [9].

In this work, we demonstrate the importance of considering the class imbalance in supervised anomaly detection when using node classification techniques. This work is motivated by the increased usage of anomaly detection datasets in node classification works [8,16,17], as well as many recent works ignoring the class imbalance issue. We employ various node embedding methods including task-dependent and structural network embedding for generating node representations, while employing several classifiers for the downstream node classification task. We evaluate the performance of various approaches in two different settings namely *balanced* and *unbalanced* which are defined based on the distribution of the minority anomalous class. We investigate various evaluation metrics in either setting to thoroughly contrast the characteristics of different settings and provide recommendations for choosing practical strategies when dealing with unbalanced datasets.

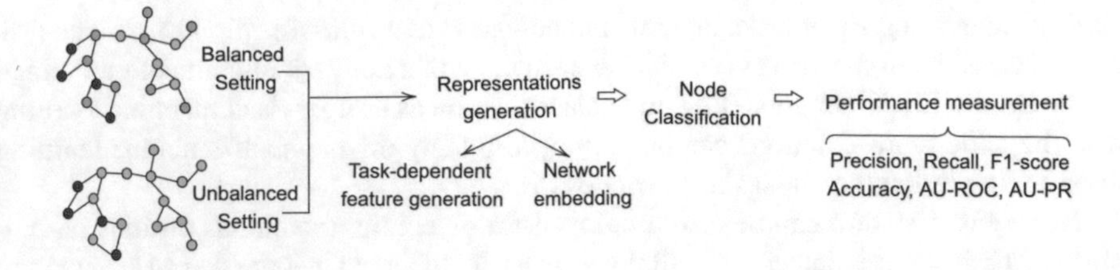

Fig. 1. Overview of the experimental pipeline. Red and green nodes respectively denote anomalous and normal nodes, while the nodes that are not included in the balanced setting classification are colored in gray.

2 Background

Network Embedding. Shallow embedding methods are basically similar to lookup tables where the vector representations of a node is found by its ID [2]. A popular category of these methods is the random walk-based approach where the embeddings are learned in an unsupervised manner. Particularly, node2vec [5], as a pioneering state-of-the-art, employs biased random walks which are fed into skip-gram architecture [12] to generate node embeddings. On the other hand, RiWalk [11] aims to learn structural node representations by decoupling identification of node roles from embedding generation procedure. RiWalk first assigns new labels to the nodes based on their structural roles. After that, it employs random walk generation and skip-gram model for extracting final embeddings. For a supervised problem such as node classification, these methods first learn the embeddings in an unsupervised way and then train a supervised classifier in the embedding space.

Anomaly Detection via Node Classification. Node classification is employed by many related researches for anomaly detection in different applications. In this regard, there are several methods that leverage heuristic approaches for generating task-dependent node representations tailored to a special application. Specifically, *REV2* [8] focuses on predicting fraudulent users on rating platforms. It considers a rating network as a signed graph and generate node representation vectors through an iterative process while incorporating behavioral attributes of the users. In cryptocurrency networks, [3,4] produce features for the Ethereum accounts based on their transaction histories, while [1,13,16] consider the Bitcoin transactions network and aim to detect instances of anomalous activities such as phishing, money laundering, and ransomware attacks.

Imbalanced Classification. Class imbalance problem refers to the situation where one of the classes have significantly less number of instances than the others [9]. Traditionally, there are three flavors of methods intending to tackle the class imbalance problem namely data-level, algorithm-level, and hybrid [9]. Data-level methods adjust the class imbalance by under- or over-sampling [9]. Algorithm-level methods assign different mis-classificaion penalties to different classes aiming to increase the importance of the minority samples [19]. Finally, the hybrid methods combines ideas from data-level as well as algorithm-level approaches to propose countermeasures against imbalance class issue [7].

3 Methodology

We focus on assessment of leveraging node classification approaches for anomaly detection in real-world networks. Given a large network of entities, where only a small portion of them are labelled as being associated with anomalous activities, the goal is to evaluate the performance of different node representations working jointly with classification methods for detection of the anomalous instances. In addition, we are interested in investigating the effectiveness of different evaluation metrics in demonstrating the performance of different approaches. We basically consider two different settings, namely *balanced* and *unbalanced*, for applying node classification. In the balanced setting, we focus on the classification of a balanced set of instances including all anomalous nodes and an equal number of normal nodes. In contrast, in the unbalanced setting, the classification is applied to all the available nodes of the networks. The performance of the node classification task in each setting is inspected based on different evaluation metrics and several recommendations from a practitioner's perspective are provided. As shown in Fig. 1, for both of the the setting, different sets of experimental assessments are performed. Mainly, we consider generating node representations by a task-dependent or a general network embedding method. The generated node representations are then employed by several classifiers for detection of anomalous instances. At the end, the performance of the downstream task are evaluated with different performance metrics.

3.1 Datasets

We assess six real-world datasets whose statistics are presented in Table 1.

- ***Bitcoin:*** is proposed by Weber et al. [16] and consists of roughly $200K$ transactions of which 2% are associated with malicious activities (e.g. ransomware or scams), 21% are considered to be from normal participants, and the rest of the transactions are unlabelled. The dataset consists of a set of carefully designed features for nodes.
- ***Ethereum:*** consists of the transaction histories of different Ethereum accounts of which 1165 accounts are associated with anomalous activities such as phishing or scam [17]. There are no ground truth labels for the rest of the accounts .
- ***OTC:*** the participants of Bitcoin transactions network use the OTC platform for rating other users [8]. The ratings range between -10 and 10, and are assigned as the edge weights. The nodes are labelled based on the platform's founder rating values. The founder together with all the other nodes he has positively rated (≥ 5) are labelled as normal, while the nodes that are rated highly negative (≤ -5) by the founder or any other normal node is labelled as fraudster.
- ***Alpha:*** the user-to-user trust network of Bitcoin Alpha platform is provided by Kumar et al. [8]. The rating and labels are set in the same way as the OTC dataset.
- ***Amazon:*** consists of a bipartite user-to-product rating network [8]. The users with at least 50 votes, where the fraction of helpful-to-total votes is ≥ 0.75, are considered normal users. In a similar way, fraudulent users are those who receives at least 50 votes where their fraction of helpful-to-total votes is ≤ 0.25.

Table 1. Statistics of the datasets. **Imbalance ratio* is defined as the number of anomalous nodes divided by the total number of nodes.

Dataset	Nodes	Edges	Average degree	Anomalous nodes	Imbalance ratio*	Network category
Bitcoin	203,769	234,355	1.3002	4,545	0.0223	cryptocurrency
Ethereum	2,973,489	13,551,303	9.1148	1,165	0.0004	cryptocurrency
Alpha	3,783	14,124	7.4671	102	0.0270	cryptocurrency rating
OTC	5,881	21,492	7.3090	180	0.0306	cryptocurrency rating
Amazon	330,317	560,804	3.3956	241	0.0007	general rating
Epinions	195,805	4,835,208	42.7914	1,013	0.0052	general rating

Table 2. Performance evaluation on cryptocurrency networks.

			Balanced setting						*Unbalanced* setting					
	Algorithm		Precision	Recall	F_1	Accuracy	AU-ROC	AU-PR	Precision	Recall	F_1	Accuracy	AU-ROC	AU-PR
Bitcoin	*RF*	Weber et al. [16]	0.987	0.845	0.911	0.911	0.982	0.986	0.923	0.687	0.788	0.921	0.955	0.888
		node2vec	0.804	0.378	0.513	0.617	0.684	0.745	0.710	0.119	0.203	0.806	0.602	0.379
		RiWalk	0.632	0.364	0.453	0.540	0.582	0.647	0.728	0.108	0.189	0.806	0.635	0.397
	LR	Weber et al. [16]	0.918	0.906	0.912	0.906	0.972	0.974	0.824	0.700	0.757	0.905	0.940	0.777
		node2vec	0.694	0.285	0.399	0.548	0.579	0.662	0.813	0.150	0.254	0.816	0.652	0.472
		RiWalk	0.666	0.261	0.367	0.533	0.574	0.643	0.735	0.082	0.146	0.801	0.635	0.393
	MLP	Weber et al. [16]	0.934	0.861	0.896	0.892	0.961	0.966	0.730	0.855	0.787	0.902	0.949	0.873
		node2vec	0.732	0.435	0.480	0.573	0.596	0.665	0.315	0.540	0.393	0.646	0.624	0.387
		RiWalk	0.745	0.245	0.358	0.540	0.571	0.641	0.360	0.472	0.408	0.706	0.634	0.391
Ethereum	*RF*	trans2vec	0.888	0.973	0.928	0.920	0.961	0.958	0.110	0.002	0.004	0.956	0.829	0.198
		node2vec	0.891	0.972	0.930	0.922	0.967	0.964	0.488	0.039	0.072	0.956	0.815	0.212
		RiWalk	0.918	0.932	0.925	0.920	0.934	0.939	0.100	0.000	0.001	0.956	0.805	0.178
	LR	trans2vec	0.911	0.910	0.911	0.905	0.958	0.958	0.276	0.073	0.115	0.951	0.783	0.178
		node2vec	0.915	0.932	0.923	0.918	0.964	0.959	0.286	0.146	0.192	0.946	0.799	0.173
		RiWalk	0.921	0.852	0.885	0.882	0.932	0.932	0.120	0.009	0.017	0.951	0.726	0.095
	MLP	trans2vec	0.890	0.970	0.928	0.921	0.955	0.952	0.097	0.730	0.170	0.676	0.761	0.164
		node2vec	0.887	0.973	0.928	0.920	0.955	0.953	0.103	0.836	0.183	0.665	0.797	0.160
		RiWalk	0.912	0.899	0.905	0.900	0.913	0.911	0.093	0.761	0.165	0.655	0.723	0.100

- ***Epinions:*** specifies the rating network of users to posts on Epinions platform where the rating values vary in $[1, 6]$ [8]. For defining the ground truth labels, a user-to-user trust network is constructed in which each user is labelled as fraudulent in the case that its total trust rate is ≤ -10, and if its total trust rate is $\geq +10$, the user is considered as normal [8].

4 Experimental Assessment

We aim to provide a comparative perspective of *balanced* and *unbalanced* setting that are currently used by the practitioners for the anomaly detection problem.

4.1 Evaluation Setup

We focused on assessing the performance of node classification approaches for anomaly detection in two distinct settings.

Table 3. Performance evaluation on rating cryptocurrency networks.

			Balanced setting						*Unbalanced* setting					
	Algorithm		Precision	Recall	F_1	Accuracy	AU-ROC	AU-PR	Precision	Recall	F_1	Accuracy	AU-ROC	AU-PR
Alpha	*RF*	REV2	0.827	0.645	0.702	0.771	0.818	0.723	0.802	0.659	0.698	0.746	0.815	0.701
		node2vec	0.672	0.629	0.644	0.698	0.761	0.710	1.000	0.000	0.000	0.973	0.571	0.031
		RiWalk	0.681	0.650	0.660	0.708	0.782	0.718	1.000	0.000	0.000	0.973	0.602	0.042
	LR	REV2	0.785	0.671	0.708	0.758	0.833	0.762	0.611	0.532	0.569	0.879	0.788	0.711
		node2vec	0.714	0.622	0.658	0.715	0.804	0.745	1.000	0.000	0.000	0.973	0.611	0.051
		RiWalk	0.698	0.590	0.630	0.701	0.798	0.759	1.000	0.000	0.000	0.973	0.596	0.047
	MLP	REV2	0.883	0.594	0.708	0.756	0.820	0.646	0.689	0.469	0.558	0.847	0.887	0.464
		node2vec	0.690	0.743	0.698	0.724	0.736	0.657	0.039	0.641	0.073	0.552	0.600	0.036
		RiWalk	0.706	0.740	0.701	0.724	0.747	0.671	0.054	0.519	0.085	0.630	0.572	0.039
OTC	*RF*	REV2	0.818	0.858	0.838	0.823	0.897	0.835	0.631	0.672	0.651	0.879	0.922	0.598
		node2vec	0.779	0.853	0.823	0.789	0.829	0.828	0.600	0.021	0.041	0.969	0.740	0.109
		RiWalk	0.776	0.816	0.794	0.758	0.801	0.802	0.800	0.027	0.052	0.970	0.737	0.110
	LR	REV2	0.802	0.787	0.794	0.801	0.873	0.842	0.596	0.643	0.619	0.833	0.855	0.687
		node2vec	0.790	0.760	0.772	0.742	0.819	0.838	0.100	0.003	0.006	0.969	0.740	0.111
		RiWalk	0.767	0.744	0.753	0.721	0.788	0.805	1.000	0.000	0.000	0.969	0.728	0.106
	MLP	REV2	0.835	0.722	0.774	0.788	0.766	0.718	0.704	0.568	0.635	0.877	0.852	0.547
		node2vec	0.778	0.795	0.774	0.736	0.715	0.728	0.067	0.660	0.122	0.703	0.675	0.056
		RiWalk	0.749	0.767	0.724	0.692	0.642	0.689	0.062	0.619	0.112	0.694	0.648	0.056

Balanced Setting. In this setting, we intended to eliminate the dataset imbalance through under-sampling of the normal nodes with the goal of attaining a roughly balanced dataset for the classification task. We considered all the available nodes in node representation learning procedure. Then, we preserved all the anomalous nodes, and randomly selected similar number of normal nodes to produce the set of node features that is used by the classifier. We repeated the random selection procedure of the normal nodes in 10 different runs and reported the average performance.

Unbalanced Setting. This is the conventional setting used in majority of the node classification approaches where all nodes are considered in representation learning as well as in the classification procedure. The advantage of this setting is that all available information is exploited for the classification. However, since the distribution of the classes in datasets containing anomalous instances is highly skewed, the performance of the classifier can be severely damaged.

In each of the two different settings, first the node representations were generated by either a task-dependent method (Weber et al. [16] for Bitcoin, trans2vec [17] for Ethereum, or REV2 [8] for cryptocurrency rating and rating network) or a general network embedding technique (node2vec [5] or RiWalk [11]). Then, the node representations were leveraged by a binary classifier (such as Random Forest (RF), Logistic Regression (LR), or Multi-Layer Perceptron (MLP)) for the detection of the anomalous nodes. The details are elaborated as follows.

- ***Task-dependent node representation approaches***: the feature sets are specifically designed for each task based on the application as well as the underlying network.
  - **Weber et al.** [16]. For the Bitcoin network dataset, we considered the feature set provided by Weber et al. [16] as the node representations for the downstream classification task. The proposed feature set consisted of 94 features

Table 4. Performance evaluation on rating networks. **NC* denotes that the node representation approach did not converged in reasonable time (the time limit was two days).

			Balanced Setting						*Unbalanced* Setting					
	Algorithm		Precision	Recall	F_1	Accuracy	AU-ROC	AU-PR	Precision	Recall	F_1	Accuracy	AU-ROC	AU-PR
Amazon	*RF*	REV2	0.801	0.811	0.806	0.804	0.854	0.678	0.587	0.623	0.604	0.919	0.921	0.536
		node2vec	0.501	0.509	0.502	0.484	0.470	0.491	1.000	0.000	0.000	0.999	0.480	0.001
		RiWalk	0.551	0.606	0.576	0.538	0.545	0.562	1.000	0.000	0.000	0.999	0.519	0.001
	LR	REV2	0.802	0.799	0.800	0.807	0.847	0.841	0.647	0.591	0.618	0.927	0.914	0.657
		node2vec	0.519	1.000	0.683	0.519	0.500	0.759	1.000	0.000	0.000	0.999	0.500	0.500
		RiWalk	0.519	1.000	0.683	0.519	0.500	0.759	1.000	0.000	0.000	0.999	0.500	0.500
	MLP	REV2	0.816	0.678	0.739	0.769	0.811	0.789	0.624	0.657	0.633	0.837	0.807	0.769
		node2vec	0.558	0.462	0.440	0.534	0.464	0.505	0.001	0.711	0.002	0.345	0.473	0.001
		RiWalk	0.589	0.606	0.543	0.549	0.487	0.515	0.001	0.687	0.002	0.367	0.485	0.001
Epinions	*RF*	REV2	0.821	0.768	0.794	0.863	0.877	0.873	0.576	0.498	0.534	0.944	0.896	0.655
		node2vec	NC*	NC	NC	NC	NC	NC	NC	NC	NC	NC	NC	NC
		RiWalk	NC	NC	NC	NC	NC	NC	NC	NC	NC	NC	NC	NC
	LR	REV2	0.769	0.758	0.764	0.841	0.856	0.857	0.503	0.434	0.466	0.895	0.887	0.699
		node2vec	NC	NC	NC	NC	NC	NC	NC	NC	NC	NC	NC	NC
		RiWalk	NC	NC	NC	NC	NC	NC	NC	NC	NC	NC	NC	NC
	MLP	REV2	0.834	0.713	0.771	0.827	0.892	0.773	0.648	0.597	0.621	0.898	0.896	0.624
		node2vec	NC	NC	NC	NC	NC	NC	NC	NC	NC	NC	NC	NC
		RiWalk	NC	NC	NC	NC	NC	NC	NC	NC	NC	NC	NC	NC

Table 5. Correlation and average difference of various evaluation metrics in balanced and unbalanced setting. For computing performance differences, the value of a metric in unbalanced setting is deducted from its counterpart in balanced setting.

	Precision	Recall	F_1	Accuracy	AU-ROC	AU-PR
Correlation	−0.216	0.244	0.421	0.017	0.785	0.366
Average Difference	0.072	0.388	0.386	−0.165	0.061	0.355

which expressed the local information about each transaction node (e.g., the timestamp and transaction fee) and 72 aggregated features gained by aggregating information from direct neighbors of each node.

- **Trans2vec** [17]. For Ethereum transaction network, we employed trans2vec [17] for generating the node representations for Ethereum accounts. Trans2vec is a random walk-based node embedding method that exploited the timestamp and amount of transactions in edge weight generation which were then used to direct the selection of nodes in random walks. The parameters of the trans2vec were set inline with those of node2vec for a fair comparison.
- **REV2** [8]. Regarding the rating platforms, namely cryptocurrency rating and general rating, we adopted the node representation learning approach proposed by Kumar et al. [8] which is the state-of-the-art approach for the task of detecting fraudulent users in rating platforms. REV2 leveraged an iterative process where the network information as well as the behavioral properties were used for generating the node representations.

• ***Network embedding approaches***. Another group of node representation methods that we have exploited for gaining node features were two state-of-the-art shallow network embedding methods namely node2vec and RiWalk that are random walk-based methods generating node representations in an unsupervised manner. In line with [5,11], we set the parameters of node2vec and RiWalk as follows: walk length $l = 5$, number of walks per node $r = 20$, embedding size $d = 64$, context size $k = 10$, and $p = 0.25$ and $q = 4$ for better exploitation of the structural equivalency of the nodes.

- ***Binary classifiers***. Gaining the node representations from the aforementioned approaches, we utilized three different classifiers for the downstream node classification task. We tested Random Forest (RF), Logistic Regression (LR), and Multi-Layer Perceptron (MLP) as our supervised classifiers. The implementation details of the classifiers are as follows: RF with number of estimator = 50, maximum number of features = 10, and maximum depth = 5, and LR with *L1* regularization were implemented using Scikit-learn Python package. The MLP was implemented in PyTorch with three layers and ReLU activation function.

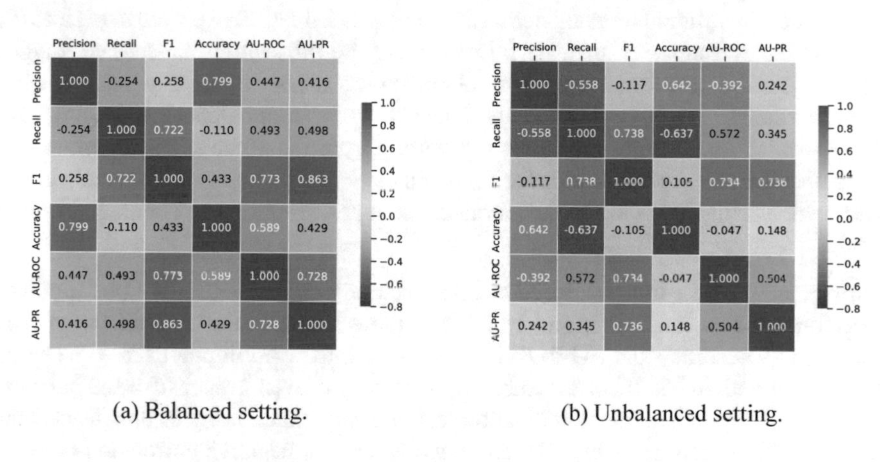

(a) Balanced setting. (b) Unbalanced setting.

Fig. 2. Correlation of different evaluation metrics in balanced or unbalanced setting.

4.2 Results Analysis and Recommendations

Considering the experimental results of node classification in two different setting, in Tables 2, 3, and 4, we can make several observations. First, it can be observed that task-dependent approaches mostly generate more efficient representations that achieve higher performance of node classification in both balanced and unbalanced setting, since these methods incorporate the extra information available in the dataset as edge weights or edge timestamps. While the general network embedding methods (i.e., node2vec and RiWalk) do not generally leverage edge features. Task-dependent methods define their sets of features according to the intrinsic characteristics of the networks and their contents. Therefore, although their application is tailored to a specific task, they mostly show higher performance on their specific platform compared to more general approaches.

Moreover, we observe that different performance metrics demonstrate different characteristics in balanced and unbalanced setting. For example, while high value of accuracy is observed in both settings, other performance measures, like precision and recall, that are more commonly adopted in datasets with imbalance issue, tell a different story. Specially in unbalanced setting, methods can predict all instances as negative and still achieve a high accuracy value. However, as detection of the positive instances in the anomaly detection task is of great importance, this scenario is not appealing and infers the necessity of better performance metrics.

We have also presented the correlation and average difference (i.e., $bal.\ perf. - unbal.\ perf.$) of each performance metric in the two settings in Table 5, and the correlation of various performance metric in either balanced or unbalanced setting in Fig. 2. Considering the results, we make the following recommendations.

Recommendation 1: The evaluation setting when detecting anomalies by a node classification approach is important and the performance in the balanced setting does not correlate closely with the performance in unbalanced setting as shown in Table 5. Different approaches should be evaluated in the setting that they are expected to perform in, otherwise the evaluation of their performance may be misleading. Indeed, we suggest to investigate the performance of node classification tasks in balanced setting, while evaluating the performance of anomaly detection tasks in unbalanced setting; or considering evaluation in both setting for a general purpose approach. This is inline with the fact that the real-world datasets for the anomaly detection tasks are extremely unbalanced, which is highly important to be considered when designing an anomaly detection method.

Recommendation 2: In unbalanced setting, accuracy and AU-ROC are not appropriate performance metrics. As shown in Table 2, Table 3, and Table 4, classification tasks achieved high accuracy and AU-ROC in most of the cases. Additionally, in some cases, the accuracy and AU-ROC of the unbalanced setting is even higher than the balanced setting, while obviously the classification task is more challenging in unbalanced setting due to high data imbalance. Notably, we observe a negative correlation between AU-ROC and precision in the unbalanced setting, shown in Fig. 2 and Table 5. It can be inducted that precision and recall are more informative performance metrics for the imbalance classification problem because they focus on the prediction of the positive instances. However, the negative correlation of AU-ROC and precision in the unbalanced setting infers that AU-ROC is not a reliable metric when the dataset is highly unbalanced.

Recommendation 3: For anomaly detection task, the balanced setting overestimates the performance as shown in Table 2, Table 3, and Table 4. Evaluating an approach in balanced setting although may results in better performance of the classification task, is far from the actual setting in anomaly detection problems where normal samples outnumber anomalous ones. In addition, the difference of the performance between these two setting could be considerable as shown in Table 5. Hence, performance metrics in balanced and unbalanced setting do not always show high positive correlation, which implies that reaching good performance in balanced setting does not necessarily results in good performance in the unbalanced setting as well.

5 Conclusion

In this paper, we explored the exploitation of node classification methods for anomaly detection in the context of real-world large graphs. We assessed the performance of node classification for an anomaly detection task in balanced and unbalanced setting. We investigated different performance metrics in our evaluation and showed that the tasks, settings, and performance metrics should be selected in accordance with the intrinsic characteristics of the datasets and use cases. As a future direction, it is interesting to investigate the performance of other representation learning methods. Important examples of these methods include graph neural networks that jointly generate node representations and classify the instances, and are trained end-to-end. Considering the training of these methods being aware of the dataset imbalance is another challenging future work.

References

1. Akcora, C.G., Li, Y., Gel, Y.R., Kantarcioglu, M.: Bitcoinheist: topological data analysis for ransomware prediction on the bitcoin blockchain. In: Proceedings of the Twenty-Ninth International Joint Conference on Artificial Intelligence (IJCAI) (2020)
2. Chami, I., Abu-El-Haija, S., Perozzi, B., Ré, C., Murphy, K.: Machine learning on graphs: a model and comprehensive taxonomy. arXiv preprint arXiv:2005.03675 (2020)
3. Chen, W., Guo, X., Chen, Z., Zheng, Z., Lu, Y.: Phishing scam detection on ethereum: towards financial security for blockchain ecosystem. In: International Joint Conferences on Artificial Intelligence Organization, pp. 4506–4512 (2020)
4. Farrugia, S., Ellul, J., Azzopardi, G.: Detection of illicit accounts over the ethereum blockchain. Expert Syst. Appl. **150**, 113318 (2020)
5. Grover, A., Leskovec, J.: node2vec: scalable feature learning for networks. In: Proceedings of the 22nd ACM SIGKDD International Conference on Knowledge Discovery and Data Mining, pp. 855–864 (2016)
6. Hamilton, W.L., Ying, R., Leskovec, J.: Inductive representation learning on large graphs. In: Proceedings of the 31st International Conference on Neural Information Processing Systems, pp. 1025–1035 (2017)
7. He, H., Garcia, E.A.: Learning from imbalanced data. IEEE Trans. Knowl. Data Eng. **21**(9), 1263–1284 (2009)
8. Kumar, S., Hooi, B., Makhija, D., Kumar, M., Faloutsos, C., Subrahmanian, V.: Rev2: fraudulent user prediction in rating platforms. In: Proceedings of the Eleventh ACM International Conference on Web Search and Data Mining, pp. 333–341 (2018)
9. Longadge, R., Dongre, S.: Class imbalance problem in data mining review. arXiv preprint arXiv:1305.1707 (2013)
10. Ma, X., et al.: A comprehensive survey on graph anomaly detection with deep learning. IEEE Trans. Knowl. Data Eng. (2021)
11. Ma, X., Qin, G., Qiu, Z., Zheng, M., Wang, Z.: Riwalk: fast structural node embedding via role identification. In: 2019 IEEE International Conference on Data Mining (ICDM), pp. 478–487. IEEE (2019)
12. Mikolov, T., Sutskever, I., Chen, K., Corrado, G.S., Dean, J.: Distributed representations of words and phrases and their compositionality. Adv. Neural. Inf. Process. Syst. **26**, 3111–3119 (2013)

13. Nerurkar, P., Bhirud, S., Patel, D., Ludinard, R., Busnel, Y., Kumari, S.: Supervised learning model for identifying illegal activities in Bitcoin. Appl. Intell. **51**(6), 3824–3843 (2020). https://doi.org/10.1007/s10489-020-02048-w
14. Poursafaei, F., Rabbany, R., Zilic, Z.: SigTran: signature vectors for detecting illicit activities in blockchain transaction networks. In: Karlapalem, K., et al. (eds.) PAKDD 2021. LNCS (LNAI), vol. 12712, pp. 27–39. Springer, Cham (2021). https://doi.org/10.1007/978-3-030-75762-5_3
15. Sakr, S., et al.: The future is big graphs: a community view on graph processing systems. Commun. ACM **64**(9), 62–71 (2021)
16. Weber, M., et al.: Anti-money laundering in bitcoin: Experimenting with graph convolutional networks for financial forensics. arXiv preprint arXiv:1908.02591 (2019)
17. Wu, J., et al.: Who are the phishers? phishing scam detection on ethereum via network embedding. IEEE Trans. Syst. Man Cybern. Syst. (2020)
18. Zhao, T., Zhang, X., Wang, S.: Graphsmote: imbalanced node classification on graphs with graph neural networks. In: Proceedings of the 14th ACM International Conference on Web Search and Data Mining, pp. 833–841 (2021)
19. Zhou, Z.H., Liu, X.Y.: Training cost-sensitive neural networks with methods addressing the class imbalance problem. IEEE Trans. Knowl. Data Eng. **18**(1), 63–77 (2005)

Agriculture Stimulates Chinese GDP: A Machine Learning Approach

Nan Zhenghan and Omar Dib(✉)

Department of Computer Science, Wenzhou-Kean University, Wenzhou, China
odib@kean.edu

Abstract. GDP is a convincing indicator measuring comprehensive national strength. It is crucial since the industrial structure, living standards, and consumption level are closely related to GDP. In recent years, the Chinese GDP has maintained rapid growth. Admittedly, the contribution of agriculture to GPD is gradually decreasing. As the foundation of life, the structure of agricultural production still needs to be improved. Therefore, this paper applies machine learning skills to investigate how to improve the agricultural production structure to promote GDP. A total of 47 agricultural products were selected and analyzed. We extracted the production data from 1980 to 2018. K-means clustering model was used to group products into several clusters. The Holt-winters model predicts the following year's production of the different agriculture products to simulate next year's GDP. The linear regression model quantifies the relationship between clusters and GDP. Based on that relationship, we provide suggestions on stimulating GDP growth. For assessment, both linear regression and neural network models are used to simulate the GDP after considering the recommendations. Results show that the proposed approach offers relevant recommendations to stimulate the Chinese GDP based on the agriculture data.

Keywords: GDP · Agriculture · Decision making · Machine learning · K-means clustering · Holt-winters · Linear regression · Neural networks

1 Introduction

According to Yi (2021), GDP (Gross Domestic Product) is the total monetary or market value of all the finished goods and services produced within a country's borders in a specific period. It's important because it reflects a country's economic strength and international status. In recent years, the Chinese GDP has maintained rapid growth. However, the development of the Chinese economy still has some issues. The industry is still the central pillar of Chinese economic growth, but this tendency violates sustainable development. Therefore, transformation to the service industry is the ultimate goal of development. Although the influence of agriculture on Chinese GDP decreases because of advanced technology, it is vital as the foundation of life. Therefore, this paper will analyze the impact of different agricultural products on GDP. Based on the analysis, a suggestion on how to modify the structure of agricultural production will be made.

© The Author(s), under exclusive license to Springer Nature Switzerland AG 2023
L. C. Tang and H. Wang (Eds.): BDET 2022, LNDECT 150, pp. 21–36, 2023.
https://doi.org/10.1007/978-3-031-17548-0_3

Many researchers have already studied the relationship between GDP and agriculture, education, and other factors. Thereby governments can stimulate the GDP growth based on those relationships. In (Kira et al. 2013), the authors adopted the Keynes model to observe the economic development trend and gave some suggestions on stimulating the GDP. But they focused on the macroeconomy field. Therefore, they didn't cluster different factors to explore the impact of other clusters on GDP. In Ifa & Guetat's (2018) paper, the authors applied the Auto-Regressive Distributive Lags (ARDL) approach to explore the relationship between public spending on education and GDP per capita. Nevertheless, they only described the positive and negative correlation of these two factors. They didn't find the formula to manifest the quantified relationship between education and GDP. Furthermore, with only positive and negative correlations, it isn't easy to give concrete suggestions on stimulating GDP growth.

Unlike previous papers, many others ignored other factors and only focused on the relationship between agriculture and GDP. Hussain (2011) used the linear model, Ordinary Least Square (OLS), to research the contribution of agriculture growth rate towards GDP growth rate. As can be noticed, the authors assumed linearity of the variables, which may not always be the case in practice. Indeed, the relationship between variables might either be linear or non-linear. In Azlan's (2020) paper, they compared Artificial Neural Network (ANN) and ordinary least squares (OLS) model performance on predicting the stock. The results show that ANN has a more accurate prediction because it can describe the non-linear relationship.

Predicting each cluster's output is necessary to simulate next year's GDP. Yoon (2020) applied advanced machine learning skills to predict GDP under different real-world settings. They used a gradient boosting model and a random forest model. In another research, when Shih (2019) compared machine learning skills and time series, they indicated that time series methods have the most negligible value of the error measures than machine learning skills. Thus, time series methods can give a better accuracy when predicting.

The previous papers didn't research the influence of factors on GDP in the microdomain. Even if some studied the microdomain, they didn't quantify the relationship between those factors and GDP. Some papers quantified this relationship, but they omitted the non-linear relationship. As for the time analysis, some articles didn't apply time series methods to provide better prediction accuracy.

Machine learning algorithms have also been used to extract knowledge from supervised and unsupervised data in various fields. For example, Liu (2021) investigates the roles and interactions of gene variants, using machine learning (ML) and big data analysis to discover the potential autism spectrum disorder. Dou (2021) proposed an auto-machine learning approach for predicting the risk of progression to active tuberculosis based on Its association with host genetic variations. Kong (2021) developed a model-free, machine-learning-based solution to exploit reservoir computing in the normal functioning regime with a chaotic attractor.

This paper applies K-means Xu (2019) clustering to observe the impact of different clusters in agriculture on GDP. Holt-Winters model is used to predict the output of clusters due to its reliability. After that, the linear regression methods will quantify the relationship between those clusters and GDP. Based on the relationship between clusters and GDP, a suggestion on stimulating GDP growth is made. Finally, both linear regression and neural networks Aggarwal (2018) are used to assess the performance of the proposed approach.

2 Methodology

This section presents the methodology adopted to apply machine learning skills to explore the improvement of the agricultural production structure to stimulate GDP. We summarize the workflow of the study in Fig. 1 and elaborate on the steps as follows. Firstly, the data is retrieved from two sources. One is China's statistical yearbook (2020): The GDP data from 1980 to 2018. The other is the National Bureau of Statistics of China, Agriculture part: 1. The output of Tea and Fruits 2. The production of Major Farm Products 3. The output of Livestock Products 4. The output of Aquatic Products (From 1980 to 2018).

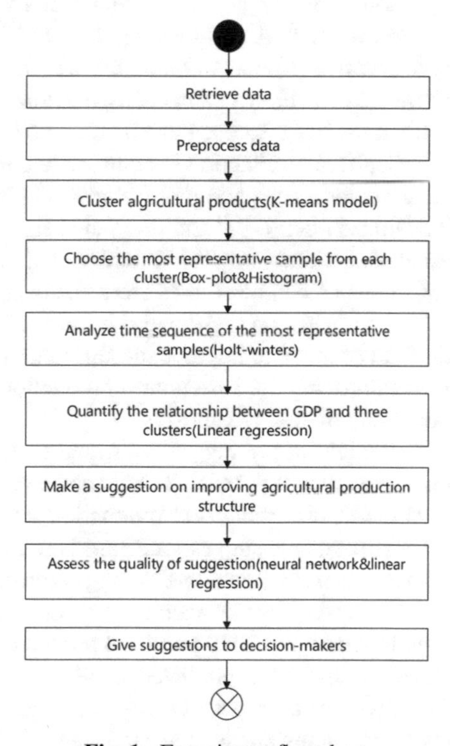

Fig. 1. Experiment flowchart.

Secondly, the data is preprocessed before applying machine learning skills. Mainly, the data is cleaned and combined. The reason is that a mass of production data is not continuous from 1980 to 2018. Many data were primarily missed before 2000 because of the backward traceability system. Thus, those products that don't have continuous data are omitted. Then the remaining 47 agricultural production data from 1980 to 2018 is combined with the GDP data from 1980 to 2018. In this way, the relationship between GDP and agricultural production can be studied. Thirdly, those agricultural products need to be clustered. In this paper, there are 47 samples, which constitutes the real data. However, this paper handles the agricultural production structure on a microscopic level. If the whole agriculture data is not divided into sub-parts, this paper can only study the macroscopic relationship between agriculture and GDP. Therefore, it's important to cluster the agricultural data so that the robustness of the internal structure of agriculture can be closely researched.

Fourthly, the most representative samples need to be selected from each cluster. The reason is that studying the impact of each agricultural product in the same cluster on GDP can't reflect the superiority of clustering. Clustering has already extracted the commonality of different products and integrated them into the same cluster based on the commonality. That's why repeatedly studying the impact of those similar products on GDP is meaningless and duplicated. Due to the commonality of various agricultural products in the same cluster, each cluster can be regarded as the minimum unit. However, each cluster is the integration of many products. Hence, it's vital to find a product representing the whole cluster to study the relationship between each cluster and GDP.

Fifthly, with the representative products obtained from the fourth step, the time sequence is analyzed. There are two aims of interpreting the time sequence of each usual product. One is that the trend of time sequence can be obtained. A recommendation is proposed to improve the agricultural production structure to stimulate the Chinese GDP. The other is to assist the simulation part, and more details will be shown in the eighth step. Sixthly, the relationship between GDP and three clusters is quantified. This step aims to find which clusters can promote the GDP and which clusters can impair the GDP. Also, the degree of influence can be studied from the weight of each cluster. By considering the effect of each cluster on GDP and the trend of each cluster obtained from step5, the suggestion can be made to improve the agricultural production structure. Seventhly, a suggestion on improving the agricultural production structure is proposed based on the influence and trend mentioned above.

Eighthly, due to the uncertainty of the suggestion's quality, it's significant to assess it. This paper assumes that the current year is 2014. Therefore, a suggestion was proposed in 2014 to improve the 2015 agricultural production structure. 2015 predicted production data would be a base for the suggestion. Based on 2015 predicted data, this paper simulates that the government implements the suggestion on 2015 predicted data. Then the expected production can either be increased or decreased. Then, the simulated GDP can be obtained based on the 2015 modified production. Also, the predicted GDP can be obtained based on the 2015 predicted GDP. Finally, the predicted 2015 GDP and simulated 2015 GDP are compared to assess the quality of the suggestion. After the evaluation, only a convincing suggestion can be applied in real-world settings. Ninthly, once the validity of the suggestions is confirmed, they are communicated with the decision-makers to improve the agricultural production structure.

3 Results and Analysis

3.1 K-Means Clustering

There is a total of 47 agricultural products. It's meaningless and impossible to research the impact of each product on the GDP. In addition, all of them belong to agriculture, so there must be some similarity among them. That's why K-means clustering is applied to first group them into various clusters. The value k should be determined before clustering; however, there is no k value. Therefore, k was set to be 2, 3, 4 and 5 roughly.

After observing the result of clustering under these four values, there can be a basic idea about the optimal value of k. Figure 2 presents the results under each case. The results can be evaluated externally from two indicators. One is the internal compactness, and the other is the external degree of separation. More specifically, compactness measures whether the sample points in a cluster are compact enough, such as the average distance to the cluster's center, variance, etc. The degree of separation measures whether the sample is far enough away from other clusters. From these two perspectives, both k = 2 and 3 gave acceptable results. To specify the value of k, these two techniques are involved in determining the k value. They are elbow plot Dinov (2018) and average silhouette scores. Figure 3 demonstrates the elbow plot. Smaller than the optimal number of clusters, the increase of k substantially increases the compactness of each cluster. Thus, the SSE

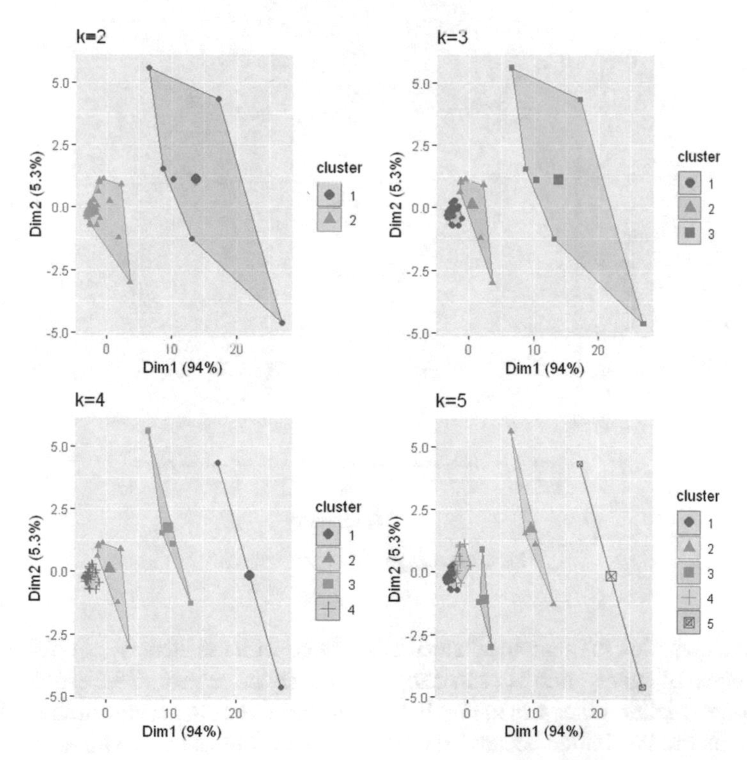

Fig. 2. Clustering under k = 2, 3, 4, 5 respectively.

drops quickly. However, when k reaches the optimal number, the expansion of k hardly impacts SSE. Therefore, the K value corresponding to this elbow is the actual clustering number.

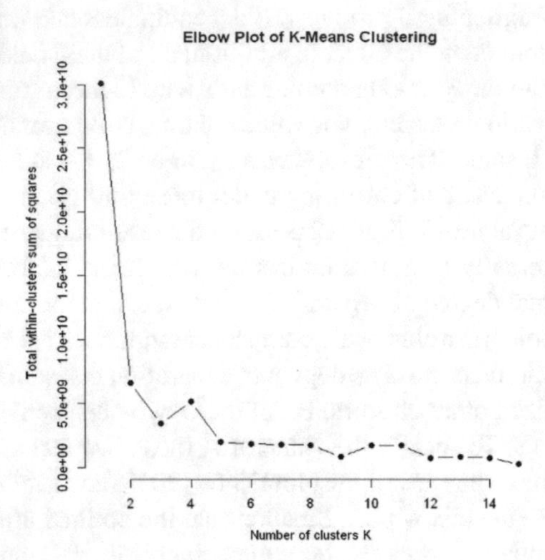

Fig. 3. Elbow plot.

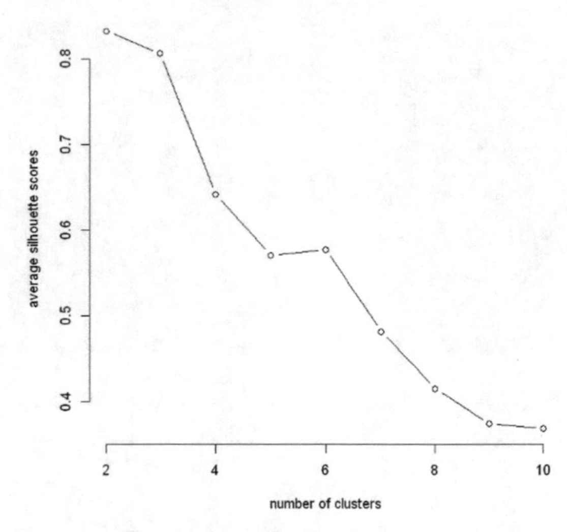

Fig. 4. Average silhouette scores.

In the elbow plot, the sample clustering is refined with the increase of cluster number K. The degree of aggregation of each cluster gradually increases, so the error square and SSE become smaller. When k is In Fig. 3, k = 2 is the actual clustering number. However, the elbow method is limited because it only considers compactness without considering the external degree of separation. Hence, Fig. 4 depicts the other approach, named average silhouette scores. The silhouette score (Shahapure, 2020) is $S = (b-a)/max(a,$

b), where a is the average distance between one sample and other samples in the same cluster, b is the average distance between the sample and all the samples in the nearest group. Here, a is the compactness above, b is the degree of separation. The smaller a is, the higher compactness is. The larger b is, the higher the degree of separation is. Therefore, k will obtain the optimal value when the S is close to 1. In Fig. 4, k = 2 is the optimal value, the same as the elbow method. To sum up, k = 2 is the real clustering number. Although the products have already been clustered wonderfully, cluster2 contains 41 samples, much more than cluster1. For the convenience of research, cluster2 was visualized to see if it could reasonably be divided into several clusters. Figure 5 demonstrates the histogram of cluster2.

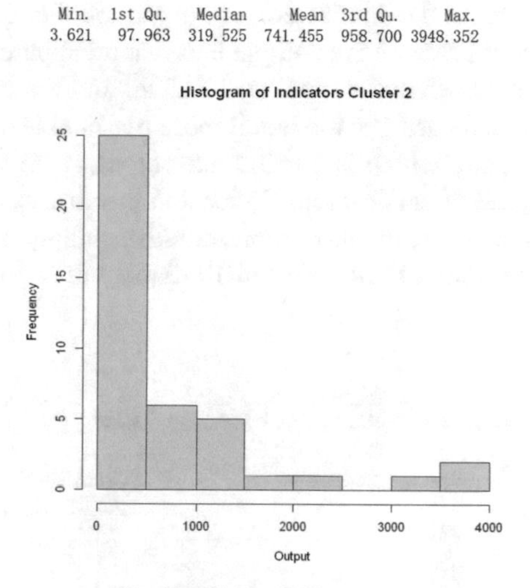

Fig. 5. Cluster2 histogram.

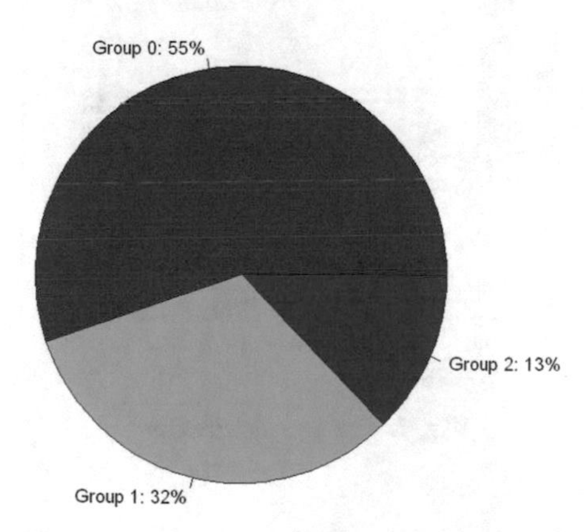

Fig. 6. Composition of agricultural products.

Here, the mean average annual output is 741.455. More than half are smaller than the mean value. Therefore, a threshold was set to separate initial cluster2 into two groups.

Those samples which are smaller than 700 will be group0. The remaining part in cluster2 is called group1. Cluster1 will be group2. Ultimately, three groups are generated for the convenience of research. Figure 6 shows the pie chart of these groups.

3.2 Clusters Analysis

After grouping all products, we select the most representative sample from each group. It is not relevant to explore the influence of each product on the GDP. Thus, the relationship between these three groups and GDP is studied. All three groups will be executed in the same manner. Here, group0 is taken as an example to demonstrate the complete process. The most representative is either the median or the mean based on the distribution of the group. Figure 7 describes the distribution of the group0. For the group0, the median was chosen as the most representative sample because an apparent right-skewed trend was found. Mode is smaller than the median, 137.526, and the median is smaller than the mean, 184.943. In this regard, the median is more reasonable to be the representative sample. Hence, the median, which is persimmon output, represents group0. Figure 8 presents the time sequence of persimmon. Notice that the abscissa begins from 0, which is 1980. With the same process, the most representative sample of group1 is apple, and garden fruit represents group2. Figures 9 and 10 display their time sequence of them, respectively.

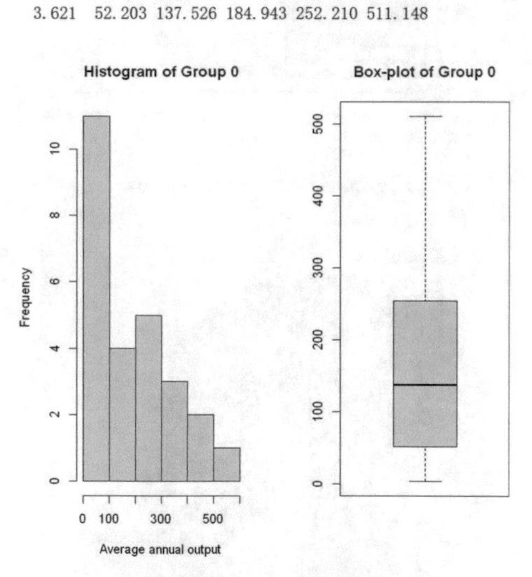

Fig. 7. Group0 distribution.

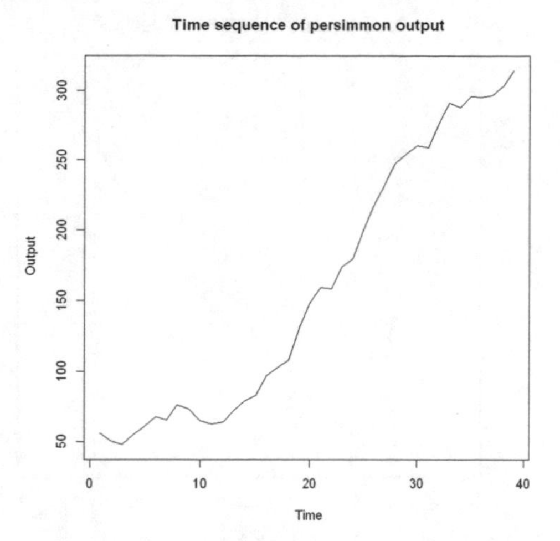

Fig. 8. Persimmon time sequence diagram.

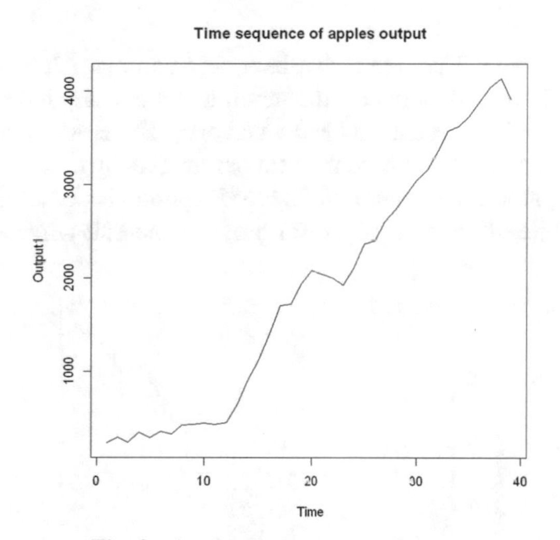

Fig. 9. Apple time sequence diagram.

3.3 Time Analysis

This paper focuses on improving the structure of agricultural production to promote GDP. To provide valid suggestions, it's essential to observe the trend of each cluster. Figures 8, 9, and 10 show the time sequence diagrams of samples. From the diagrams, the shape of the curve is quite straightforward. There is a clear upward trend and no seasonality. Thus, the holt-winters model can probably give a satisfying prediction. Actually, the holt model is competent to conduct the prediction. The holt-winters model is applied here because there might be some hidden seasonality that can be observed Canela (2019). That's why this section will apply the holt-winters model to predict the future trend of those representative samples. Take group0, for example; there are three

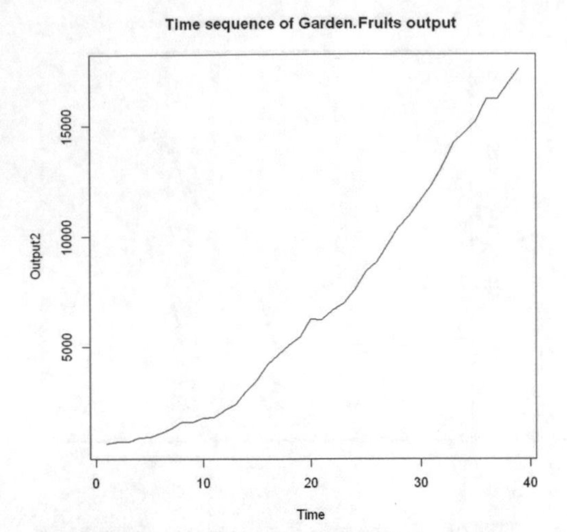

Fig. 10. Garden fruits time sequence diagram.

parameters in this model. They are alpha, beta, and gamma. Alpha is the level that is the weight of new information, beta is the trend, and gamma is the seasonality. Lower weights give less weight to recent data and vice versa. The Holt-winters function in the R language will automatically choose the best parameters by minimizing AIC and BIC values. Eventually, alpha = 1, beta = 0.2824653, gamma = FALSE. Gamma is false means that there is no seasonality. Figure 11 presents the holt-winters fitting.

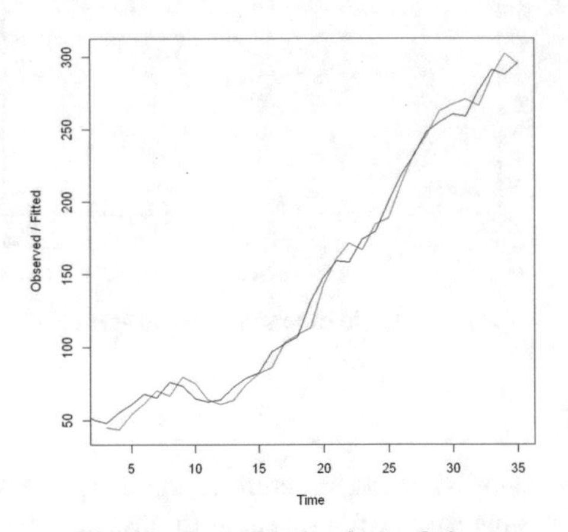

Fig. 11. Group0 holt-winters fitting.

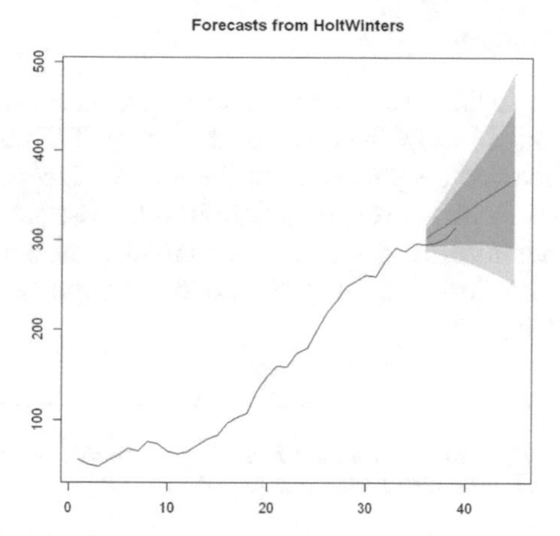

Fig. 12. Group0 forecasts from holt-winters.

	ME	RMSE	MAE	MPE	MAPE	MASE	ACF1
Training set	1.376459	7.941914	6.576597	2.065224	5.929615	0.7512071	0.02586935

Fig. 13. Group0 holt-winters fitness.

The red line is the fitting line, and the black line is the actual line. Figure 13 presents the fitness of the holt-winters model. Here are two significant indicators which can evaluate the fitness of the model. One is MAPE (mean absolute percentage error, which is approximately 6, meaning that the forecast is off by 6% on average. The other is the RMSE (root mean square error) which reflects the model's fitness. Here RMSE is 8. Furthermore, Fig. 11 shows that the predicted value is consistent with reality. After selecting the best parameters, Fig. 12 demonstrates the prediction. In this figure, holt-winters started prediction from 2015 to 2018. The red line is the actual value. The blue line is the predicted value. Dark grey is an 80% confidence interval, and light grey is a 95% confidence interval. In the worst case, the actual is still located in the dark grey part. Thus, it's a relevant prediction. It's not necessary to predict more than one year because the output of agricultural products is primarily determined by the annual policy and the previous year's output. Moreover, this paper aims to give governments some suggestions on improving the agricultural product structure. Incredibly, the advice can be valid for a long time because too many dynamic factors influence the output. When focusing on 2015, the actual output is 294.96, and the predicted one is 302.9005. As mentioned before, the predicted one belongs to an 80% confidence interval, which is a good prediction. After all, a slight modification of the policy can have a crucial impact on agriculture. The other two samples are executed similarly so that the result will be displayed directly. Likewise, for group1, the actual value of 2015 is 3889.9, and the predicted one is 3858.523.

3.4 Quantification Analysis

Before making the suggestion, how each group of products affects GDP should be explored. Figures 14 and 15 show the result of the fitting. The number of * represents the performance of each feature prediction in the model. Except for intercept, all the model shows a perfect performance of prediction. Besides, R-squared is 0.98, presenting a prominent fitness. Therefore, the linear regression model can explain their relationship well. Here is the concrete formula: GDP = −2616.6 * Group0 − 252.4 * Group1 + 151.1 * Group2 − 51558.2.

```
Call:
lm(formula = GDP ~ Output.of.Persimmons.10000.tons. + Output.of.Apples.10000.tons. +
    Output.of.Garden.Fruits.10000.tons., data = df2)

Coefficients:
                     (Intercept)     Output.of.Persimmons.10000.tons.
                         51558.2                              -2616.6
    Output.of.Apples.10000.tons.  Output.of.Garden.Fruits.10000.tons.
                          -252.4                                151.1
```

Fig. 14. Linear regression fitting.

```
Call:
lm(formula = GDP ~ Output.of.Persimmons.10000.tons. + Output.of.Apples.10000.tons. +
    Output.of.Garden.Fruits.10000.tons., data = df2)

Residuals:
   Min     1Q Median     3Q    Max
-75882 -27163   7302  25507  56726

Coefficients:
                                     Estimate Std. Error t value Pr(>|t|)
(Intercept)                          51558.199  18716.749   2.755  0.00926 **
Output.of.Persimmons.10000.tons.     -2616.608    400.035  -6.541 1.51e-07 ***
Output.of.Apples.10000.tons.          -252.433     30.916  -8.165 1.28e-09 ***
Output.of.Garden.Fruits.10000.tons.    151.105      8.749  17.271  < 2e-16 ***
---
Signif. codes:  0 '***' 0.001 '**' 0.01 '*' 0.05 '.' 0.1 ' ' 1

Residual standard error: 36860 on 35 degrees of freedom
Multiple R-squared:  0.9823,    Adjusted R-squared:  0.9808
F-statistic: 648.7 on 3 and 35 DF,  p-value: < 2.2e-16
```

Fig. 15. Linear regression fitting evaluation.

3.5 Decision Making

According to the formula at the end of Sect. 3.4, it was found that both group0 and group1 impair the GDP, whereas group2 can promote GDP. What's more, group0 has the most significant impact on GDP. Considering the output increase from a macro-perspective, next year's output is probably higher than the previous year. This paper will simulate 2014. Suppose this year is 2014, and the suggestion was proposed in 2014. The aim of the suggestion is to improve the agricultural product structure in 2015. Hence, the recommendation is that governments should slow down the output growth of group0 and group1 and promote the output growth of group2 in 2015. To get the growth of output, 2013 output should be obtained. Then the development of 2014 to 2015 can be estimated from 2013 to 2014.

Table 1. 2013–2015 real, predicted, and simulated output

Year	Output of group0 (10000.tons.)	The output of group1 (10000.tons.)	The output of group2 (10000.tons.)
2013(real)	287.87	3629.81	14675.17
2014(real)–2013(real)	7.69	105.58	496.19
2014(real)	295.56	3735.39	15171.36
2015(real)	294.96	3889.9	16200.91
2015(predicted)	302.9	3858.5	15827.66
2015(simulated)–2014(real)	4.44	54.61	928.64
2015(simulated)	300	3790	16100

Table 1 shows the actual output of 2013 and 2014. The difference between 2014 and 2013 of groups 0, 1, and 2 are displayed on the third row. Based on the discrepancy and strategy above, the growth of output from 2014 to 2015 was displayed on the seventh row. Finally, the specific suggestion is that the average production of group0 should be controlled to 300, the average group1 result should be 3790, and the average group2 output should 16100.

3.6 Decisions Evaluation

In this section, the validity of the suggestion will be estimated by comparing actual, simulated, and predicted GDP. Then, both neural network and linear regression will be applied to confirm the validity. Although the linear regression in Sect. 3.4 demonstrates exciting results, there exists a possibility that the relationship between agricultural products and GDP is non-linear. Figure 16 shows the fitted neural network.

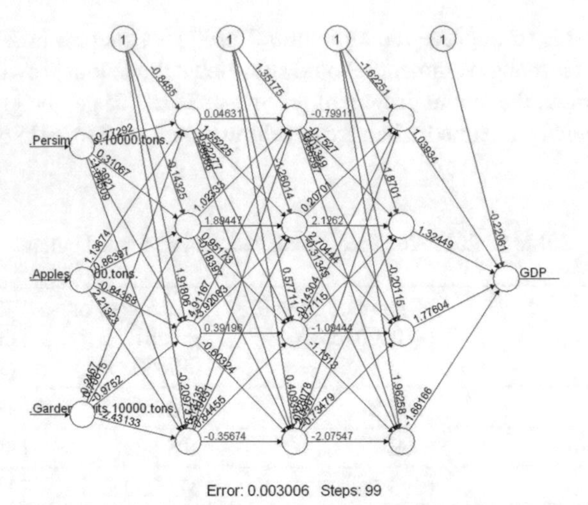

Fig. 16. Neural network

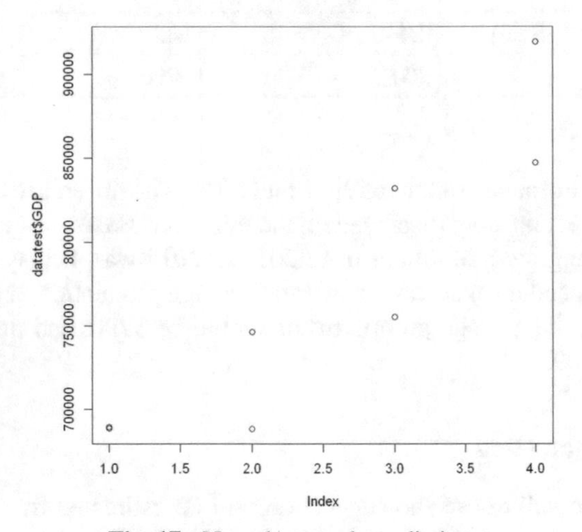

Fig. 17. Neural network prediction.

The number of hidden layers is 3, and there are four neurons in each hidden layer. The r-squared value determines the configuration. R-squared can be maximized, which is 0.5257 under this configuration. After establishing the model, Fig. 17 shows the result of the prediction. There are four years in this model which are 2015, 2016, 2017, and 2018. The red points are the predicted values, and the black points are the actual value. Although the predictions are not so accurate in the last three years, the forecast and the actual value of 2015 almost overlap. As mentioned before, this paper only needs to predict one year GDP. After predicting the 2015 GDP, the GDP after governments adopted the suggestion will be simulated. Linear regression model and neural network are applied to simulate the 2015 GDP. Table 2 shows the result.

Table 2. Simulated GDP

Year	Output of group0 (10000.tons.)	The output of group1 (10000.tons.)	Output of group2 (10000.tons.)	GDP
2015(simulated by Linear regression)	300	3790	16100	699209.5
2015(predicted by Linear regression)	302.9	3858.5	15827.66	641487.4
2015(simulated by Neural network)	300	3790	16100	705348.0
2015(predicted by Neural network)	302.9	3858.5	15827.66	694861.8
2015(real)	294.96	3889.9	16200.91	688858.2

Table 2 conveys two pieces of information. First, Table 2 reveals that the neural network performed better than linear regression while predicting the 2015 GDP. The other is that both simulated results are higher than the real one. That is to say; the suggestion is convincing and valid.

4 Conclusions and Future Works

This paper focuses on providing and verifying valid suggestions about improving agricultural production so that it can stimulate the Chinese GDP. Four machine learning skills involving K-means clustering, Holt-winters, Linear regression, and Neural network are applied throughout the whole process. The result shows that the 2015 simulated GDP is higher than the 2015 predicted GDP after adopting the suggestions, and the 2015 simulated GDP is even higher than the 2015 real GDP. Because this paper assumes that an excellent agricultural production structure can stimulate GDP, the suggestions are valid in the real world. Separately, this paper only researches the data from 1980 to 2018. The reason is that many products' output data is still being collected. 2018 is the most recent year, providing the most output data. In addition, in this work, we only applied the K-means clustering because of its simplicity and low time complexity. Other clustering algorithms can be applied to improve the accuracy. In the future, the proposed methodology in this paper can be used in other fields besides agriculture. What's more, both the time analysis and clustering models can be changed to improve the accuracy. Also, the parameters of the neural network, which are the number of neurons and layers, can be tuned to a better combination. Furthermore, the most representative products can be selected more reasonably, or the methodology can be modified to decrease the impact of a limited selection of representatives. In addition, the data can also be improved in the future. Instead of using the annual production data, the monthly data can enrich the dataset, improving the prediction accuracy.

References

1. Anwar, H., Khan, A.Q.: Relationship between agriculture and GDP growth rates in Pakistan: an econometric analysis (1961–2007). Academic Research International 1, no. 2, p. 322 (2011)
2. Azlan, A., Yusof, Y., Mohsin, M.F.M.: Univariate financial time series prediction using clonal selection algorithm. Int. J. Adv. Sci. Eng. Inf. Technol. **10**(1), 151–156 (2020)
3. Yoon, J.: Forecasting of real GDP growth using machine learning models: gradient boosting and random forest approach. Comput. Econ. **57**(1), 247–265 (2021). https://doi.org/10.1007/s10614-020-10054-w
4. Shih, H., Rajendran, S.: Comparison of time series methods and machine learning algorithms for forecasting Taiwan blood services foundation's blood supply. J. Healthcare Eng. 2019 (2019)
5. Kira, A.R.: The factors affecting Gross Domestic Product (GDP) in developing countries: the case of Tanzania (2013)
6. Ifa, A., Guetat, I.: Does public expenditure on education promote Tunisian and Moroccan GDP per capita? ARDL approach. J. Finan. Data Sci. **4**(4), 234–246 (2018)
7. Liu, Z., et al.: Machine learning approaches to investigate the relationship between genetic factors and autism spectrum disorder. In: 2021 The 4th International Conference on Machine Learning and Machine Intelligence, pp. 164–171 (2021)
8. Dou, W., et al.: An AutoML Approach for Predicting Risk of Progression to Active Tuberculosis based on Its Association with Host Genetic Variations (2021)
9. Kong, L.-W., Fan, H.-W., Grebogi, C., Lai, Y.-C.: Machine learning prediction of critical transition and system collapse. Phys. Rev. Res. **3**(1), 013090 (2021)
10. Xu, J., Lange, K.: Power k-means clustering. In: International Conference on Machine Learning, pp. 6921–6931. PMLR (2019)
11. Yi, Y.: The relationship between industrial structure and economic growth in China—an empirical study based on panel data. In: E3S Web of Conferences, vol. 275, p. 01009. EDP Sciences (2021)
12. Canela, M.Á., Alegre, I., Ibarra, A.: Holt-winters forecasting. In: Quantitative Methods for Management, pp. 121-128. Springer, Cham (2019). https://doi.org/10.1007/978-3-030-17554-2_13
13. Aggarwal, C.C.: Neural Networks and Deep Learning. Springer, vol. 10, pp. 978–3 (2018). https://doi.org/10.1007/978-3-319-94463-0
14. Dinov, Ivo D. K-Means Clustering. In: Data Science and Predictive Analytics, pp. 443-473. Springer, Cham, 2018. https://doi.org/10.1007/978-3-319-72347-1_13
15. Shahapure, K.R., Nicholas, C.: Cluster quality analysis using silhouette score. In: 2020 IEEE 7th International Conference on Data Science and Advanced Analytics (DSAA), pp. 747–748. IEEE (2020)

Short Message Service Spam Detection Using BERT

Irvan Santoso(✉)

Computer Science Department, Bina Nusantara University, Jakarta, Indonesia 11480
isantoso@binus.edu

Abstract. It is undeniable, fraud on smartphone users is still very widespread today. One of the scams that is quite common and is detrimental to smartphone users is fraud that is spread through information media such as short message services. Many users still lack knowledge which causes them to suffer losses by performing actions they do not understand. Therefore, in this study a method is proposed to detect messages sent via short message services are messages that are not dangerous or vice versa. The dataset used is 1,525 messages consisting of normal, scam, and advertisement messages. The method used in this research is classification using the BERT approach. Based on the simulation results for the classification, the average accuracy obtained is 91.13% with the highest accuracy reaching 91.33%. It can be concluded that this approach provides good value to be used as a reference in detecting incoming messages, as well as convincing smartphone users to make decisions before taking action on information.

Keywords: Spam detection · Spam classification · BERT · Short message service · Smartphone

1 Introduction

One of the problems that often occur in everyday life is the number of fraudulent actions that are often carried out by irresponsible people [1]. Fraud acts that are rampant and carried out in many ways by perpetrators are now increasing and varied [2]. One of them is through fraud by using advertising messages, spam messages, and phishing which are very disturbing to the comfort of the recipient [3]. These messages are often found in various media that are connected to devices that are used personally such as smartphones [4]. Short Message Service (SMS) is one of the media that is the most easily accessible target for fraud by fraudsters [5]. This is because the fraudsters simply make a list of numbers sequentially which can easily be automated using tools/software [6]. Although this is already widely known by the public or smartphone users in general [7]. Unfortunately, there are still many smartphone users who are still affected because they have a low level of awareness of the truth of information [8].

Based on data on fraud cases reported by news in Indonesia [9], it is noted that the trend of cybercrime has increased during the Covid-19 pandemic. During January to November 2020, there were 4,250 cybercrimes handled by the authorities. In detail,

© The Author(s), under exclusive license to Springer Nature Switzerland AG 2023
L. C. Tang and H. Wang (Eds.): BDET 2022, LNDECT 150, pp. 37–45, 2023.
https://doi.org/10.1007/978-3-031-17548-0_4

the cases consisted of defamation dominating with 1,581 cases; fraud 1,158 cases; and illegal access 267 cases. Then from the data presented, since 2020, there have been 649 reports of fraud and 39 times of data theft. There were also 18 complaints of hacking of electronic systems. This resulted in a total of 15,367 complaints with a total loss of Rp 1.23 trillion. This problem is also reinforced by data obtained from related parties handling this case, where from all online fraud reports from 2016 to September 2020, there were around 7,047 online fraud cases reported. On average, there are 1,409 cases of online fraud each year.

Judging from these data, with the increasing number of crime cases that occur every year, a way is needed to be able to detect messages that are considered fraud or not. Therefore, in this study, a method for detecting messages using Natural Language Processing (NLP) [10] will be proposed. NLP is a approach that is often used in various types of research involving sentence processing. This approach can process data automatically by using the specified model in processing words or sentences. In its development, messages will be classified into three forms, namely messages that are fraudulent; advertising messages; and messages that are of a normal nature or are not considered to be fraudulent or advertising categories. The proposed method for classifying is using the BERT Model approach, which is one of the best current classification models.

2 Related Works

In 2018, Devlin et al., [11] conducted a study by introducing a new representation model called Bidirectional Encoder Representations from Transformer (BERT). BERT is designed to be able to carry out an in-depth, two-way training process of unlabeled text by combining all contexts in layers. Contributions made from this research, among others, demonstrate the importance of two-way pre-training for language representations and demonstrate that pre-trained representations reduce the need for many specially engineered or heavily engineered architectures.

This is different from the research conducted by Radford et al., [12] which used a one-way language model for pre-training. BERT is the first fine-tuning-based representation model to achieve good performance on a large range of tasks at the sentence level and token level outperforming many custom architectures. In addition, the data used in this research is data from BookCorpus which contains eight hundred million data and English Wikipedia which contains two billion five hundred million data.

Research on BERT has also begun to develop because the process has reached an optimal level and, in the end, many researchers working in optimization or similar fields to find the best accuracy, such as Liu et al., [13] have conducted a study entitled "RoBERTa: A Robustly Optimized BERT Pretraining Approach" which aims to measure the impact of many parameters and various forms and amounts of data. The data used is BookCorpus data which contains eight hundred million data and English Wikipedia which contains two billion five hundred million data with a total of 16 GB. The contribution given from this research is the formation of the process needed to determine the appropriate alternative to the BERT training design. This determination results in better downstream task completion than other models. In addition, Liu et al., stated that the use of more data or more data in pre-training can improve performance than downstream tasks.

Lan et al., [14] also conducted a similar study entitled "ALBERT: A Lite BERT For Self-Supervised Learning of Language Representations" which aims to increase the speed of training using BERT. In his research, A Lite BERT (ALBERT) which has been designed has an architecture with fewer parameters when compared to the architecture owned by BERT. This is what causes ALBERT to reduce the time needed to do pre-training. The contribution given from this research is optimization of factorized embedding parameterization, cross-layer parameter sharing, and inter-sentence coherence loss. In addition, the data used in this study also uses BookCorpus and English Wikipedia which have been used in previous BERT research.

Wang et al., [15] conducted a study on BERT with the title "StructBERT: Incorporating Language Structures into Pre-Training for Deep Language Understanding" which aims to increase accuracy in sentence understanding. This is done by proposing a new structural pre-training that extends BERT to include structural purpose words and sentence structural objectives to improve language structure in contextual representation. This allows StructBERT to explicitly model the structure of a language by forcing it to reconstruct the correct order of words and sentences for correct predictions. The data used in this study is also the same as the previous researchers, namely BookCorpus and English Wikipedia which have a total of 16 GB.

Munikar et al., [16] also conducted research using BERT in conducting sentiment analysis with the research title "Fine-grained Sentiment Classification using BERT". In this study, the data used is data from the Standard Sentiment Treebank (SST). In a study conducted by Munikar et al. Experiments show that the proposed BERT model yields better scores than other popular models used in sentiment analysis.

In addition, Wilie et al., [17] conducted research on the development of BERT by adopting the need for an Indonesian language structure entitled "IndoNLU: Benchmark and Resources for Evaluating Indonesian Natural Language Understanding". It is known that the structure of the Indonesian language compared to the structure of other languages has a big difference. This has an impact on the results given in performing natural language processing on various languages. The data used in this research is data collected from various sources, such as blogs, news, and websites. In this study, the results obtained are that IndoNLU can provide better accuracy in carrying out natural language processing in Indonesian and this can be a reference in future research in Indonesian.

In another study that is still related to sentiment analysis of the Indonesian language, Nugroho et al., [18] conducted a study entitled "BERT Fine-Tuning for Sentiment Analysis on Indonesian Mobile Apps Reviews". This research aims to see the sentiment value of the reviews given by users of mobile-based applications. According to them, reviews from users are one of the important things that need to be collected to develop applications to increase user satisfaction. The results obtained using the BERT model are that BERT can provide a better accuracy value for sentiment analysis compared to other commonly used models.

Azhar & Khodra [19] have conducted a similar study in Indonesian with the title "Fine-tuning Pretrained Multilingual BERT Model for Indonesian Aspect-based Sentiment Analysis". In the research conducted, there is a multilingual BERT using data on Indonesian-language reviews given to hotels. The approach taken is splitting the dataset into several small datasets. The dataset that has been formed is then re-divided into several groups based on the category of aspects it has so that it becomes a new dataset to train the sentiment classification model separately. The results obtained from this study are an increase in the accuracy of the methods used in previous studies.

Therefore, based on all the previous studies that have been discussed. It can be seen that the approach using the BERT model is better than other similar models in handling similar data. Therefore, in this study, the BERT approach will be used as the main model in classifying the SMS data that has been collected.

3 Dataset

The dataset used in this research is SMS data collected from various sources consisting of three classifications, including advertising, scam, and normal. Advertising data is data that contains information on product offerings, information on new products, information related to marketing, or similar information. Scam data is data that has invalid information and has suspicious information for which the contents cannot be accounted for. Finally, normal data is data that is not classified as a scam or advertisement and has general information or is only for notification. The amount of data obtained in data collection can be seen in Table 1.

Table 1. Total dataset

SMS classification	Total
Advertisement	182
Scam	641
Normal	702
Total data	**1,525**

Based on Table 1, the total data obtained is 1,525 with the details of advertising data as much as 182 data; scam data totaling 641 data; and the normal data are 702 data. Although the advertising data is not as much as other data, it is quite representative, because the forms of advertising data are not much different, so it is easy to learn by the developed model. As an illustration of the data obtained, Table 2 is data example of each classification.

Furthermore, this research will be conducted by dividing the existing data into 75% for training and 25% for testing so that the data used for training is 1.144 and the data used for testing is 381. This division will be carried out evenly, judging by the existing classifications.

4 Research Methods

The process that will be carried out in this research to get the right classification consists of several steps, including read dataset, normalization, and classification process. In

Table 2. Sample data for each classification

SMS classification	Sample data	Description
Advertisement	Paket internet 2GB utk 7h HANYA Rp 15Rb. Aktivasi dgn balas WOW2 ke SMS ini atau cek *363*20# dan tsel.me/hotoffers Promo sd 27 Mei	Offers about internet package
	Bingung milih paket yg pas buatmu? Coba cek *888*15# skrg juga! Siapa tau ada yang klik!	
Scam	pelanggang yth no,wa Anda M-dptkn cek 175jta pin (WHA012) silahkan cek pin anda di, bit.ly/inforesmi-33	Scam for money transfer
	N.O.Anda dpt Rp.150.000.000 dri program TOKOPEDIA thn.2021 kode pin (KM435UGD) silahkan cocokkan pin anda di; tinyurl.com/pediainfoid	
Normal	Selamat, Paket Combo Sakti 2 GB, 75 Menit Tsel, 400 SMS Tsel & Langganan Disney + Hotstar /30 hari Rp 26000 telah aktif, berlaku s/d tgl 04/07/2021 pkl. 23:59 WIB. Cek status/berhenti berlangganan melalui My Telkomsel Apps atau hub *363#. Info: 188	Notification of internet package
	Pelanggan yth, Paket Combo Sakti akan berakhir pada tanggal 04/07/2021 jam 23:59 WIB	

normalization, there are several processes carried out to ensure the data obtained is easy to use for classification. If the classification is still carried out without normalization, the results obtained will not provide good accuracy, because in the data used there are many errors in spelling and in the form of symbols that need to be filtered. These stages are shown in Fig. 1.

Based on Fig. 1, the dataset used is the data that has been described previously, scilicet data with the nature of advertising, scam, and normal. The data will be normalized by considering several things, such as the characteristics of the data and errors that appear that are not appropriate in writing a sentence or related matters. In normalization, the first thing to do is filter strings such as checking for new lines, excess whitespace, and

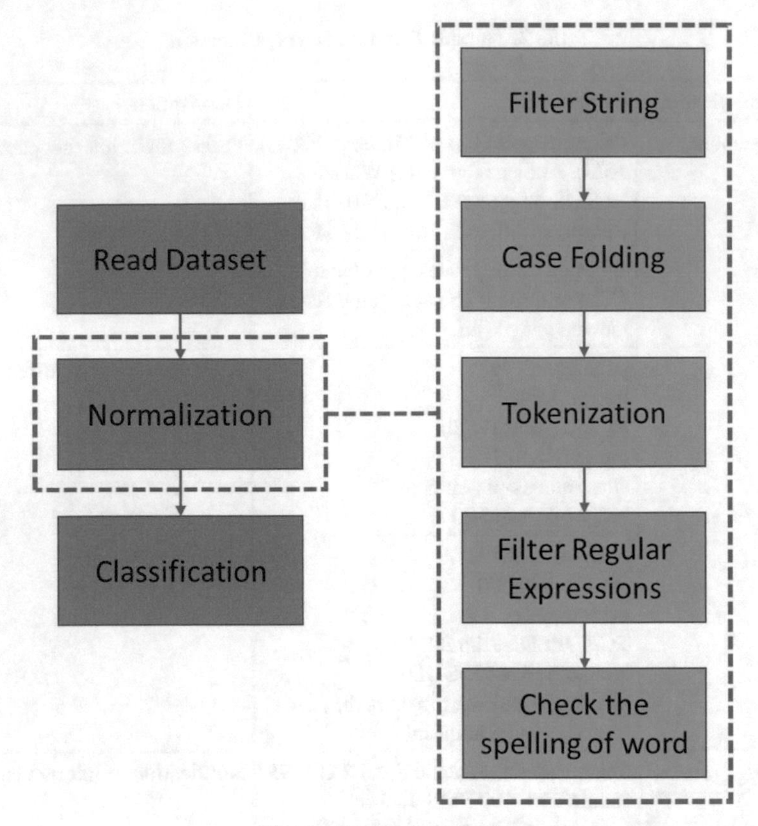

Fig. 1. Research methods

indentation. If there is an excess of those things, it will be adjusted, so there are no meaningless duplicate characters like double whitespace. In addition, sometimes strings in SMS take the form of emojis or emoticons. Therefore, in this process, emojis are also converted to strings, such as emoji ":)" to ":smile:". Furthermore, case folding will be performed to change all capital letters into lower cases to make checking easier. Then, tokenization is performed to separate the string and to check for regular expressions. Since the regular expression does not give any meaning in determining the classification, the regular expression will be omitted when checking. After the regular expression is removed, a spell check will then be performed to make the classification more accurate. Spell checking will use the Symspell [20] approach and classification will use the BERT Model approach. For instance, the example of normalization will be shown in Table 3.

Table 3. Example for normalization

SMS classification	Sample data	Result	Translation
Scam	Pelanggang yth no,wa Anda M-dptkn cek 175jta pin (WHA012) silahkan cek pin anda di, bit.ly/inforesmi-33	Pelanggan yang terhormat nomor whatsapp anda mendapatkan cek 175 juta pin wha012 silahkan cek pun anda di bitly info resmi 33	Dear customer, your whatsapp number got a check for 175 million pin wha012, please check yours at bitly official info 33

5 Results and Discussion

Based on the classification carried out on the data obtained, the BERT model approach provides a good accuracy value of more than 90%. However, to find out whether this model is valid and can produce good values continuously. It takes several trials to ensure that the values obtained continuously are not much different. To ensure that, the method is simulated up to ten times and each result is checked as shown in Table 4.

Table 4. Simulation results

Attempt	Accuracy
1st	91.27%
2nd	91.33%
3rd	91.21%
4th	91.22%
5th	90.89%
6th	90.99%
7th	91.01%
8th	91.12%
9th	91.21%
10th	91.07%
Average	**91.13%**

Based on Table 4, from the ten simulation experiments carried out, the highest accuracy that can be obtained is 91.33% in the second simulation and the average is 91.13%. When viewed from the experiments performed, the results obtained are not much different from each other and can provide a good accuracy value. For information, the data used is 25% of the overall data obtained in accordance with the previous information, scilicet 381 SMS data consisting of advertisements, scams, and normal data. In addition, the experimental results only look at the accuracy without looking at the processing time.

6 Conclusion

The rise of fraud that occurs in cyberspace threatens to cause huge losses for smartphone users. The number of users who do not pay attention to this is easily provoked into taking actions that can harm themselves. One of the most common scams is the one that is spread via SMS. Therefore, in this study a method is proposed to detect messages sent via SMS are valid messages or not scams. The approach taken in this study uses the BERT model and produces good accuracy in classifying SMS. This can be used as a reference in determining the right action for smartphone users before trusting the information obtained from SMS. In addition to future improvement, this method needs to be developed and implemented into an application that can be downloaded by smartphone users. This is to make it easier for every smartphone user to make the right decision in taking action on the information obtained.

References

1. Lana, A.: Dampak kejahatan siber terhadap teknologi informasi dan pengendalian internal. J. Econ. Soc. Educ. **1**(3), 1–13 (2021)
2. Nurse, J.R.: Cybercrime and you: how criminals attack and the human factors that they seek to exploit (2018). arXiv preprint arXiv:1811.06624
3. Ibrahim, S., Nnamani, D.I., Soyele, O.E.: An analysis of various types of cybercrime and ways to prevent them. Int. J. Educ. Soc. Sci. Res. **3**(02), 274–279 (2020)
4. Duha, N.: Short message services (SMS) fraud against mobile telephone provider consumer review from law number 8 of 1999 concerning consumer protection. J. Law Sci. **3**(1), 36–43 (2021)
5. Pervaiz, F., et al.: An assessment of SMS fraud in Pakistan. In: Proceedings of the 2nd ACM SIGCAS Conference on Computing and Sustainable Societies, pp. 195–205 (2019)
6. Dewi, F.K., Fadhlurrahman, M.M.R., Rahmanianto, M.D., Mahendra, R.: Multiclass SMS message categorization: beyond spam binary classification. In: 2017 International Conference on Advanced Computer Science and Information Systems (ICACSIS), pp. 210–215. IEEE (2017)
7. Breitinger, F., Tully-Doyle, R., Hassenfeldt, C.: A survey on smartphone user's security choices, awareness and education. Comput. Secur. **88**, 101647 (2020)
8. C. N. N. Indonesia, Polri tangani 4.250 kejahatan Siber Saat pandemi, nasional, 01-Dec-2020. https://www.cnnindonesia.com/nasional/20201201141213-12-576592/polri-tangani-4250-kejahatan-siber-saat-pandemi. Accessed 20 Dec 2021
9. Qiu, X., Sun, T., Xu, Y., Shao, Y., Dai, N., Huang, X.: Pre-trained models for natural language processing: a survey. Sci. China Technol. Sci. **63**(10), 1872–1897 (2020). https://doi.org/10.1007/s11431-020-1647-3
10. Devlin, J., Chang, M.W., Lee, K., Toutanova, K.: Bert: pre-training of deep bidirectional transformers for language understanding. arXiv preprint arXiv:1810.04805
11. Radford, A., Narasimhan, K., Salimans, T., Sutskever, I.: Improving language understanding with unsupervised learning (2018)
12. Liu, Y., et al.: Roberta: a robustly optimized bert pretraining approach (2019). arXiv preprint arXiv:1907.11692
13. Lan, Z., Chen, M., Goodman, S., Gimpel, K., Sharma, P., Soricut, R.: Albert: a lite bert for self-supervised learning of language representations (2019) arXiv:1909.11942

14. Wang, W., et al.: Structbert: incorporating language structures into pre-training for deep language understanding (2019). arXiv preprint arXiv:1908.04577
15. Munikar, M., Shakya, S., Shrestha, A.: Fine-grained sentiment classification using BERT. In: 2019 Artificial Intelligence for Transforming Business and Society (AITB), vol. 1, pp. 1–5. IEEE (2019)
16. Wilie, B., et al.: IndoNLU: benchmark and resources for evaluating Indonesian natural language understanding (2020). arXiv preprint arXiv:2009.05387
17. Nugroho, K.S., Sukmadewa, A.Y., Wuswilahaken DW, H., Bachtiar, F.A., Yudistira, N.: BERT fine-tuning for sentiment analysis on Indonesian mobile apps reviews. In: 6th International Conference on Sustainable Information Engineering and Technology, pp. 258–264 (2021)
18. Azhar, A.N., Khodra, M.L.: Fine-tuning pretrained multilingual BERT model for Indonesian aspect-based sentiment analysis. In: 2020 7th International Conference on Advance Informatics: Concepts, Theory and Applications (ICAICTA), pp. 1–6. IEEE (2020)
19. Murugan, S., Bakthavatchalam, T.A., Sankarasubbu, M.: SymSpell and LSTM based Spell-Checkers for Tamil (2020)

Big Data and Data Management

A Study on the Application of Big Data Technology in the Excavation of Intangible Cultural Resources

Min Yuan(✉)

Department of Textiles and Apparel, Shandong Vocational College of Science and Technology, Weifang 0536, China
215790831@qq.com

Abstract. Intangible cultural heritage is the core element of national and national culture. In some areas and people in China, the awareness of protection of intangible cultural heritage is relatively weak. Some traditional crafts and techniques are gradually lost due to lack of inheritance. Therefore, we urgently need to use modern technical means to bring valuable intangible cultural resources back to the public eye. The purpose of this paper is to accurately mine intangible cultural resources through big data technology. The methods used in this paper mainly include literature research method, interdisciplinary research method and statistical method. This paper builds a data analysis platform by inducing and analyzing the keywords of intangible cultural heritage, and then realizes the mining of intangible cultural resources by building a big data management platform and a big data analysis platform. Research shows that big data technology is more accurate than traditional data processing methods, and can effectively detect the public's preference for intangible cultural heritage. Big data technology can realize a benign mechanism of mutual promotion in the fields of culture, art, communication and information.

Keywords: Big data · Intangible cultural resources · Cultural inheritance · Mining application

1 Introduction

Intangible cultural heritage is a typical representative of a country's history and culture. At present, 40 projects in China have been included in the UNESCO intangible cultural heritage list. China's intangible cultural heritage ranks first in the world, representing China's 5000 year precious cultural wealth. However, people's awareness of the protection of intangible cultural assets is relatively weak. Due to the lack of inheritance, some traditional crafts and technologies are slowly disappearing from public view. With the development of Internet technology, big data technology has become a necessary means of cultural protection and inheritance. The wide application of big data technology has injected new vitality into the cultural industry and improved people's cognitive level. It is of great significance to use the big data technology of intangible cultural heritage to

© The Author(s), under exclusive license to Springer Nature Switzerland AG 2023
L. C. Tang and H. Wang (Eds.): BDET 2022, LNDECT 150, pp. 49–59, 2023.
https://doi.org/10.1007/978-3-031-17548-0_5

realize the digitization of intangible cultural heritage. People can understand and appreciate the intangible cultural heritage of the motherland anytime and anywhere on the mobile terminal. This can enhance people's sense of identity with national culture and enable them to inherit and develop traditional skills.

2 Problems Existing in the Mining of Intangible Cultural Resources by Big Data Technology

2.1 Characteristics of Big Data

It is generally believed that big data refers to the collection of massive data that cannot be acquired, stored, analyzed, transmitted and used within a generally tolerated time period using traditional information technology and software and hardware tools. Big data has the following four characteristics. One is large size. As the amount of information in the real world increases exponentially, the scale of data we have is expanding, and the unit of measuring data information is not the previous GB and TB, but PB, EB, and ZB. According to an existing research report by the Internet Data Center (IDC), in the next 10 years, the world's big data is expected to increase by 50 times, and the number of servers used to manage data warehouses will increase by 10 times on the current basis. Second, there are many modes (Variety). After entering the era of big data, there are more types of data structures than before. In addition to conventional structured data (Structured Data), there are more and more semi-structured data (Semi-structured Data) and unstructured data (Unstructured Data)). As the Internet and mobile Internet penetrate deeper into people's life, study and work, unstructured data has increased substantially, and its proportion in all data has an absolute advantage. Accompanying this is a change in the way data is linked. The third is to generate Velocity. Big data is mainly generated in the form of data flow, which has the characteristics of dynamic, fast and strong timeliness. In order to make good use of data flow, it must be within an effective time, otherwise the opportunity will be fleeting. In addition, the value and state of the data we own will change with small changes in time and space, and there will also be emergent features that change with time. Fourth, the value is huge, but the density is very low. Extracting the desired value from the data is the basic starting point of our study of the data, and it is practically impossible to determine whether the amount of data increases its value in a positive correlation. The current technology's ability to process big data increases the difficulty of obtaining value from it, making the value density still very low.

2.2 Existing Problems

An important means of protecting and developing intangible cultural heritage is to use modern information technology and network technology to construct a resource database. At present, all parts of the country are exploring the construction of intangible cultural heritage resource database through various channels. Although some preliminary achievements have been made, there is still a certain gap with the development trend of big data.

First, the number of resource databases is seriously insufficient. Compared with the number of China's intangible cultural heritage, the number of established resource databases is far from enough, and there is a significant lag. According to statistics, the intangible cultural heritage with a resource database only accounts for less than 6% of the featured collections or self-built collections of Chinese public libraries. The vast majority of them are simply links to relevant public websites via web pages. As an important force in the construction of resource database, the university library has done a lot of work, but it is still not enough in the construction of intangible cultural heritage resource database. Second, the content of the resource database is too simple. At present, the resource database is mainly established based on the list of intangible cultural heritage, most of which are just a collection of simple contents, and there are not many special research materials added. In addition, there is the problem of unbalanced content in the resource database. Affected by factors such as region, economy and culture, the resource database construction of some intangible cultural heritage projects is relatively large-scale. The opposite is true for most intangible cultural heritage items, with almost zero content. In addition, from the current overall situation of all intangible cultural heritage resource data, the total amount of information data is very limited. Third, the function of the resource database is too simple. As mentioned earlier, extracting value from data in the era of big data is the main goal of building a database, and it is not enough at present. Most of the resource databases that have been built do not have an attractive presentation interface, and the functions of the database are too simple. The technology of intangible cultural heritage resource data with website as the main carrier is backward, and the lack of personalized functions has caused serious obstacles to users. There is no information retrieval function or the function is not strong, which makes the utilization rate of resource data very low. There are almost no application functions, the result is that the added value of intangible cultural heritage resources cannot be realized in the form of network, and the channel to the industry or market is lost. Fourth, the degree of cooperation in resource database construction is low. China's intangible cultural heritage resources are rich and distributed in different regions. Due to the unbalanced attention paid by local governments, the level of economic development, and technological conditions, the construction of resource databases varies significantly. At present, the construction of different resource databases is often "separate", and there is no cooperation at all. All these result in the low overall level of resource database construction, low construction speed, and high repetition rate, resulting in a waste of manpower and material resources. Therefore, it is urgent to improve the degree of cooperation in the construction of resource database. Fifth, the utilization rate of the resource database is too low. Although some resource databases have been built, many of them are "shelved" and have not played their due role, and there have been many phenomena of "heavy construction and neglect of use".

3 Construction of Intangible Cultural Resources Mining System

3.1 Basic Forms of Intangible Cultural Heritage

Intangible cultural heritage is a cultural treasure that reflects the wisdom of the nation [1]. Intangible cultural heritage is also the cultural gene of a nation. The types of Chinese intangible cultural heritage mainly include musical instruments, arts, folk songs, festivals, Chinese calligraphy, crafts, techniques, beliefs, skills, and folk arts. Specifically include Xinjiang Uyghur Muqam Art, Qinghai Regong Art, Mongolian Khumai, Xi'an Drum Music, Korean Nongmusic Dance, Mongolian Long-tuned Folk Song, Guizhou Dong Song, Oracle Bone Inscription, Seal Carving, Chinese Paper-cut, Anhui Rice Paper, Zhejiang Longquan Celadon Technology, woodblock printing, wooden architecture, Chinese acupuncture, Guangdong Cantonese opera, Tibetan opera, Peking opera, shadow puppetry [2, 3].

3.2 The Basic Idea of Applying Big Data Analysis

Information on intangible cultural heritage can be obtained from various sources. Figure 1 shows the basic framework for applying big data to intangible cultural heritage. Museums can inquire about physical objects and valuable photographs of intangible cultural heritage [4]. These photos make the intangible cultural heritage more concrete and vivid and allow people to deeply understand history. Information on intangible cultural heritage on the Internet is mainly documentary films. Through historical recordings, people can learn more about the past and present of intangible cultural heritage. At the same time, people can also collect information about intangible cultural heritage through resource sharing and resource co-construction, and then use big data technology to gain

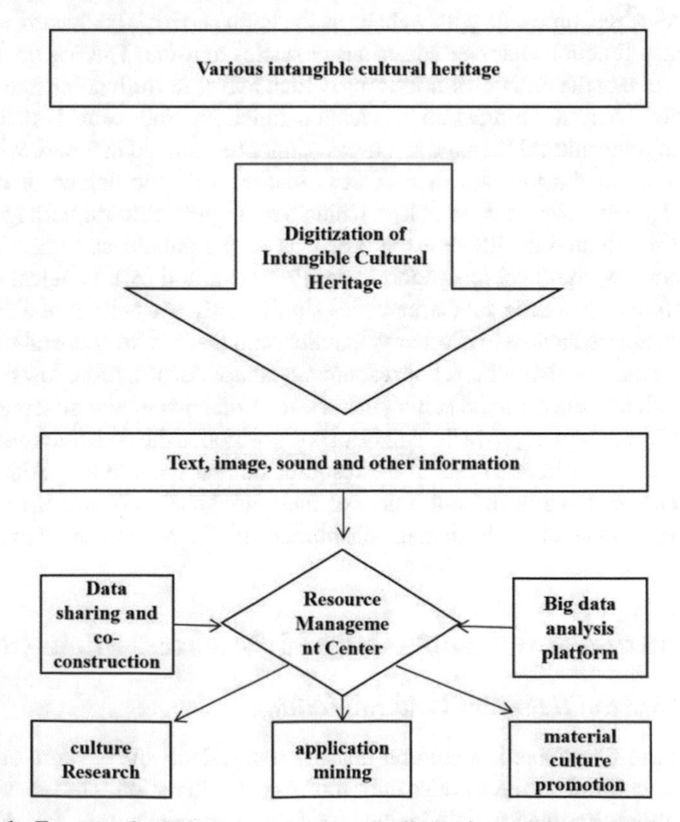

Fig. 1. Framework of big data technology applied to intangible cultural heritage

and apply intangible cultural heritage [5]. In a word, people can collect all available intangible cultural heritage information and then use modern digital technologies to convert it into graphics, images, sounds and other data information. At the same time, the aggregated intangible cultural heritage information is transmitted to the data center to realize the integration of data resources.

3.3 Construction of Data Management Platform

In building the intangible cultural heritage data management platform, we must first summarize various intangible cultural heritage data and resources, including books, collections and materials related to intangible cultural heritage. Various information needs to be aggregated and then translated into data through a digital transformation platform. Finally, the data management platform requires unified scheduling [6, 7].

After sorting the massive data, you can use mobile internet software designed to send content based on customer usage habits. Platform information can not only become important data for cultural research, but also a promotional medium for graphics, images and news. It is beneficial to increase people's sense of national identity and cultural self-confidence [8–10]. Figure 2 is the basic diagram of the Intangible Cultural Resource Data Management Platform.

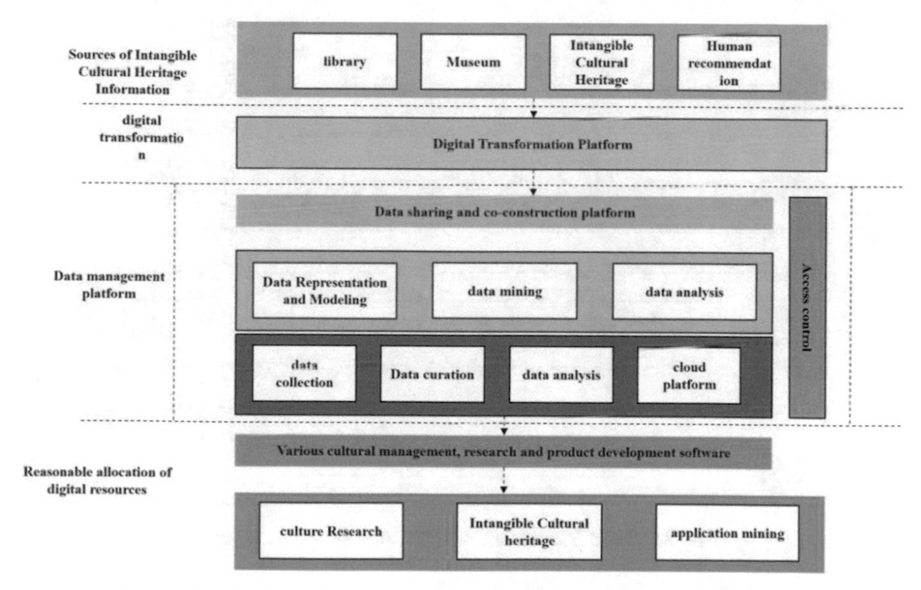

Fig. 2. Basic framework of the data management platform

3.4 Data Analysis Platform Construction

Data analysis is very important for the application and decision making of Big Data Processing technology. Data consolidation unifies and converts multiple resources into data. This includes relational database data extraction, real-time data acquisition, file data acquisition, and real-time database replication. Data can be retrieved and calculated at any time after its storage. Computing data is the key to all big data applications, including bulk computing, traffic computing, content computing and query computing. These calculations can reasonably lead information on intangible cultural heritage to meet user or market needs. Data mining is also important. After the data has been calculated, reasonable parsing is required: descriptive parsing, predictive parsing and deep parsing with data mining algorithms. Effective data computing is a prerequisite for mining applications, such as decision support. Finally, calculations and mining results must be displayed mobile or multimedia. There are different forms of exhibitions, and they are basically visualized exhibitions and interpretations of results. Data parsing and its platform structure are both challenging and important for the entire system. Data mining is also the most important part of intangible cultural assets. Figure 3 shows the basic framework for building a data parsing platform.

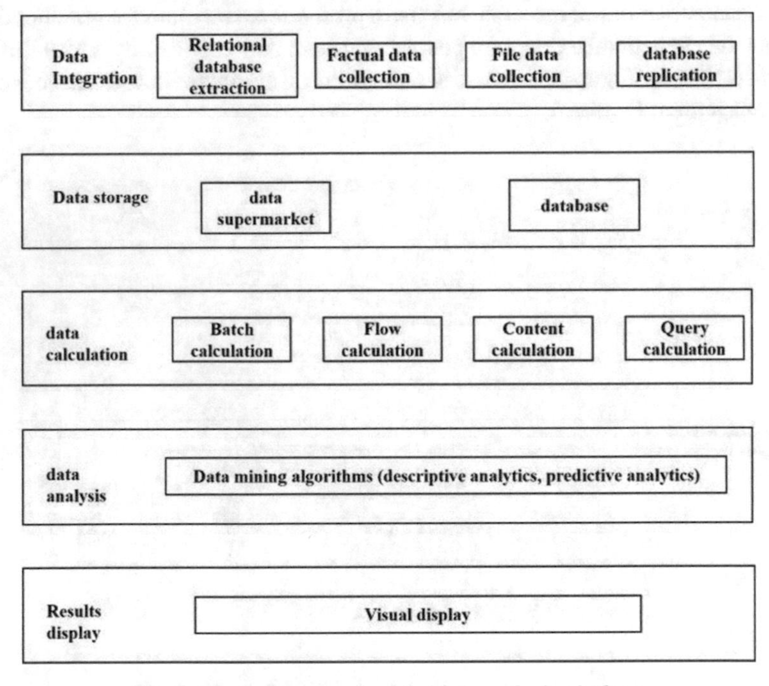

Fig. 3. Basic framework of the data analysis platform

3.5 The Specific Implementation of Intangible Cultural Resources Mining

Using big data technology to mine intangible cultural resources, we must first summarize and manage the original data. This is also a preparation for subsequent calculations. When the data is ready, the resources need to be digitized. The intangible heritage model is designed and interpreted systematically through the intangible heritage model. Through the original knowledge and features, we can make better use of these data. Figure 4 shows the specific implementation process of intangible cultural resources mining.

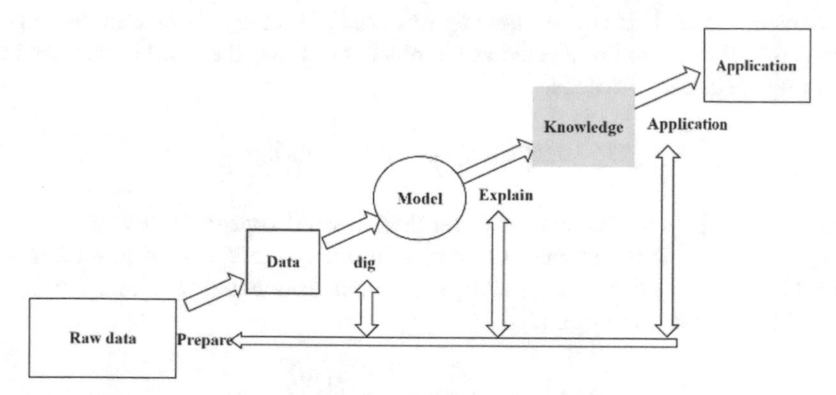

Fig. 4. The implementation process of intangible cultural resources mining

4 Data Processing

4.1 Data Collection

Information sources can take many paths in collecting data on intangible cultural resources. Information is collected and disseminated in traditional ways, such as information published by traditional media such as cultural centres, libraries, museums, TV shows, newspapers and magazines. It also disseminates and stores information through new media, such as websites, electronic databases, intangible heritage sites, Weibo, WeChat public accounts, forums and social platforms. This information has different properties of homogeneous and heterogeneous. It also has different properties of structured, unstructured or semi-structured. Therefore, it is necessary to use different data processing methods for processing. In addition, China has a variety of intangible cultural heritage resources, and each intangible cultural resource has its own unique characteristics. In addition to the relevant information obtained by the above methods, historical background, developmental genes and ethnic characteristics related to intangible cultural heritage resources should also be collected by searching historical books or related collection materials. It is necessary to collect more keywords such as dynasty, age, related persons and role in order to improve the integrity of the data.

4.2 Data Analysis

Suppose that the collection of all intangible cultural heritage we want to represent is $D = \{d1, d2, d3\ldots, dN\}$. The set (also called dictionary) of all words that appear in intangible cultural heritage is $T = \{t1, t2, t3\ldots, tN\}$.

The j-th intangible cultural heritage is represented as $dj = \{w1j, w2j, w3j\ldots wnj\}$. Where w1j represents the weight of the first word t1 in the intangible cultural heritage j.

Larger values indicate more importance. The interpretation of other vectors is similar. In order to represent the j-th intangible cultural heritage dj, the key is to calculate the value of each component in dj. If the word t1 appears in the j-th intangible cultural heritage, w1j takes the value 1. If t1 does not appear in the j-th intangible cultural heritage, its value is 0. w1j can also be selected as the number of times the word t1 appears in the j-th intangible cultural heritage.

$$TF - IDF(t_k, d_j) = TF(t_k, d_j) \bullet \log \frac{N}{n_k} \tag{1}$$

where TF (tk, dj) is the number of times the kth word appears in intangible cultural heritage j. And nk is the number of articles including the kth word in all intangible cultural heritage. The final weight of the kth word in the intangible cultural heritage j is obtained by the following formula.

$$\mathrm{w}_{k,j} = \frac{TF - IDF(t_k, d_j)}{\sqrt{\sum_{s=1}^{|T|} TF - IDF(t_s, d_j)^2}} \tag{2}$$

4.3 Find Users with Similar Interests

Let N(u) be the set of items that user u likes. N(v) is the set of items that user v likes. Then use the Jaccard formula to calculate the similarity between u and v.

$$\mathrm{w}_{k,j} = \frac{TF - IDF(t_k, d_j)}{\sqrt{\sum_{s=1}^{|T|} TF - IDF(t_s, d_j)^2}} \tag{3}$$

Cosine similarity:

$$w_{uv} = \frac{|N(u) \cap N(v)|}{|N(u) \cup N(v)|} \tag{4}$$

Suppose there are currently 4 users A, B, C, D and 5 items a, b, c, d, and e. The relationship between the user and the item (the user likes the item) is as follows:

$$w_{uv} = \frac{|N(u) \cap N(v)|}{\left|\sqrt{N(u) \times N(v)}\right|} \tag{5}$$

Data processing based on data collection, data analysis and other technologies is the key technology research on the application of big data technology to the mining of intangible cultural resources.

5 Test Analysis

In this paper, the experiment of mining intangible cultural resources is carried out through data simulation. The experiment draws the final conclusion through the process of data collection, analysis and modeling, and calculation.

5.1 Test Procedure

This work takes Chinese calligraphy, Beijing Opera and Chinese Silk in China's Intangible Cultural Heritage as examples and crawls and collects their keywords through various books, film and television works, museums, web information, and lists of Intangible Cultural Heritage. The data is converted and stored using digital conversion technology.

Table 1. Big data analysis results

Frequency term	Key words					
	History	Figure	Story	Effect	Specific works	Total
Chinese calligraphy	0.36	0.4	0.21	0.23	0.58	0.2
Peking opera	0.34	0.5	0.46	0.32	0.46	0.32
Chinese silk	0.3	0.33	0.32	0.56	0.68	0.48

Table 2. Traditional data analysis results

Frequency term	Key Words					
	History	Figure	Story	Effect	Specific works	Total
Chinese calligraphy	0.26	0.35	0.2	0.19	0.35	0.23
Peking opera	0.24	0.45	0.35	0.22	0.3	0.44
Chinesc silk	0.26	0.34	0.29	0.45	0.54	0.49

5.2 Analysis of Results

According to Table 1 and Table 2, the analysis results are obtained by binary semantic fitting. The results are shown in Fig. 5.

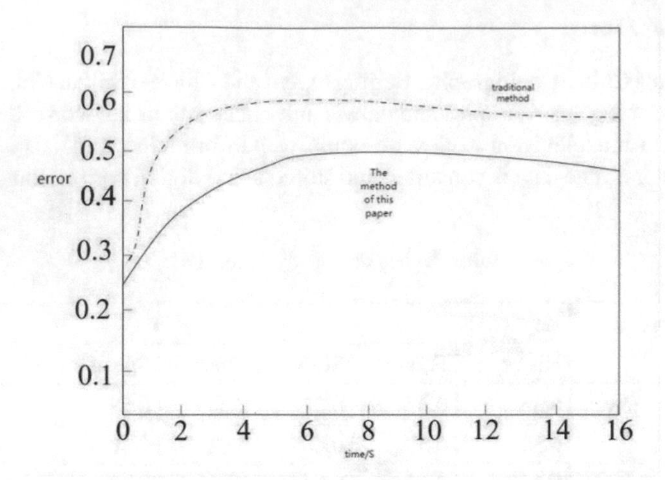

Fig. 5. Analysis results

Compared with the analysis method in this paper, the traditional data processing method has larger error and lower accuracy. It can be seen that in the art form of Chinese calligraphy, people are most interested in specific calligraphy works, so people can improve their understanding of calligraphy by printing calligraphy works collections, posters, picture books, etc. In addition, from the total number of statistics, the volunteers want to know the most about Chinese silk. Therefore, more people can learn about Chinese silk culture by further publicizing silk products and silk history.

6 Conclusion

This paper studies the application of big data technology in the mining of intangible cultural resources. Through the establishment of an intangible cultural resource management platform and an analysis platform, big data technology is used to analyze people's love for intangible cultural resources, and the validity of the data is tested. The purpose of the research is to enable China's intangible cultural heritage to be better inherited and protected. The development of intangible cultural heritage is realized through big data technology, thus forming a benign mechanism for mutual promotion in the fields of culture, art, communication and information.

Acknowledgment. This work is the gradual realization of the Heilongjiang Provincial Philosophy and Social Science Research Planning Project (No. 19YSE358) and the Shandong Provincial Social Science Planning Research Project (No. 21CPYJ79).

References

1. Zhang, X.: A preliminary study on the protection and application of historical. Architect. Cult. (02), 203–204 (2022)

2. Yin, F.: Design and application of Hebei province culture and tourism. Digit. Technol. Appl. **40**(01), 210–212 (2022)
3. Chen, X.: Using big data technology to promote the statistical research on red cultural resources in the three northern counties of Langfang. Comp. Res. Cult. Innov. **6**(01), 71–74 (2022)
4. Liang, Y., Yang, R., Xie, Y., Wang, P., Yang, A., Li, W.: Deep integration and analysis of world cultural heritage attributes based on big data. J. Remote Sens. **25**(12), 2441–2459 (2021)
5. Qin, J.: Research on the protection of historical and cultural cities in Guangzhou based on informatization and big data thinking. Innovative technology, enabling planning, and sharing the future. In: Proceedings of the 2021 China Urban Planning Informatization Annual Conference, pp. 131–137(2021)
6. Liu, Y., Zhang, Y.: Research on the construction of a big data platform for the integration of intangible cultural heritage culture in Suzhou Grand Canal. Contemp. Tour. **19**(34), 109–112 (2021)
7. Wu, X., Huang, Y.: Construction and promotion of ceramic culture inheritance platform under the background of big data. Digit. Technol. Appl. **39**(11), 193–195 (2021)
8. Xie, K.: Accurate translation and biography of Chinese culture driven by big data. J. Yanshan Univ. (Philos. Soc. Sci. Ed.) **22**(06), 12–18+28 (2021)
9. Tang, Y.: Research on the brand building of Xiashawan ancient town in the age of big data. Pop. Lit. Art (21), 49–50 (2021)
10. Xu, S.: Research on the innovation of Hui culture interactive platform in the era of big data—Taking the design of Huangshan tourism APP as an example. Green Packag. (10), 116–119 (2021)

Student Achievement Predictive Analytics Based on Educational Data Mining

Yongling Wu[1], Maria Visitacion N. Gumabay[2](✉), and Jie Wang[1]

[1] Qiannan Normal University for Nationalities, Duyun, China
[2] St. Paul University Philippines, Tuguegarao, Philippines
mvgumabay@spup.edu.ph

Abstract. This paper applies the educational data mining algorithm to the study of student achievement prediction systems using the electronic information engineering major of Qiannan Normal University for Nationalities in China as an example to design and develop a student achievement prediction system based on educational data mining. The procedure employs the simulation of electronic technology course learning performance prediction analysis as an example, using the combination of the K-means algorithm, the decision tree, the box technology, and the SVM algorithm to achieve a clustering analysis of students, course discrete processing, students' course learning performance prediction analysis, and research. By using database technology, the students' basic information, course score information, and exported forecast analysis reports can be input and stored. The correct rate of prediction and analysis of students' grades reaches 78.3% through the system's debugging test. The plan was is evaluated by the ISO/IEC 25010 software quality standard, and the system reaches the ISO/IEC 25010 software quality standard.

Keywords: Education data mining · Database · Decision tree · SVM · K-means · Student · Achievement prediction

1 Introduction

With the rapid development of information technology, network technology, and storage technology, as well as the large-scale application of the internet, cloud computing, the Internet of Things, artificial intelligence, and machine learning, a considerable amount of data resources and information have been accumulated around the world. People have entered the "era of big data." To find valuable data information required by users from massive data resources, traditional data analysis methods are combined with complex algorithms to process a large amount of data for data processing and mining technology. Specifically, data mining is for discovering previously unknown, practical, implicit, novel, and potentially valuable knowledge and rules from large amounts of data, and then shows the data to users' incomprehensible models.

In early 2020, an unprecedented COVID-19 pandemic occurred globally, which led to a requirement for schools to teach online. In addition to various online instructions, such as distance education and MOOCs, many educational data resources were generated in

© The Author(s), under exclusive license to Springer Nature Switzerland AG 2023
L. C. Tang and H. Wang (Eds.): BDET 2022, LNDECT 150, pp. 60–74, 2023.
https://doi.org/10.1007/978-3-031-17548-0_6

education called "education big data." Education big data can be divided into four types: primary data, state data, resource data, and behavioral data [1]. To discover knowledge from education big data and solve problems in education research and practice (such as assisting teaching administrators in making educational decisions, improving students' learning initiative, helping teachers improve teaching methods, etc.) under the background of education big data, the research field of education data mining has gradually formed.

Okike used clustering, classification, and visual machine learning algorithms to carry out a correlation study on learning management system (LMS) resources and students' academic performance [2]. Robbi used data mining techniques to study internet education from students in Sumatra and Java, Indonesia [3]. To find the critical moments of online learning, we use educational data mining technology to carry out research and obtain the vertical participation model as a measure to distinguish learners' performance [4]. In addition, Tong, LV, Shi, and others used educational data mining technology to research accurate early warnings of online learning, an analysis of learning behavior, supervision of the learning process, etc., and achieved corresponding research results [5–7].

Zaffar used the fast correlation filter (FCBF) feature selection method to construct a student achievement prediction model, and studied the performance of the FCBF on three different student datasets [8]. Sokkhey proposed a Michi feature selection method based on mutual information feature scoring and the chi-square algorithm [9]. Gure selected 215 male students and 2257 female students from 4422 students who participated in the PISA 2015 test to form the student dataset [10]. Fernandes analyzed and studied the performance prediction of public school students in the Federal District of Brazil from 2015 to 2016, and concluded that "performance" and "absence from class" were the most critical attributes of student performance prediction [11]. In addition, there are Apriori algorithm-based and neural networks [12], one-card data and course classification [13], one-card and multiple linear regression [14], one-card consumption data [5], and weblogs [15, 16].

2 Methodology

2.1 System Development Life Cycle

The system development life cycle (SDLC) is a phased system analysis and design method [17]. In this method, only the design and research work of the previous stage can be completed first; then, the design and research work of the next step can be continued until the system development is completed. Although each phase activity is listed separately, each phase cannot be completed in isolation. Instead, several phase activities can co-occur, and the same phase activity can occur repeatedly. The phased workflow of the development life period of the student achievement prediction system is shown in Fig. 1.

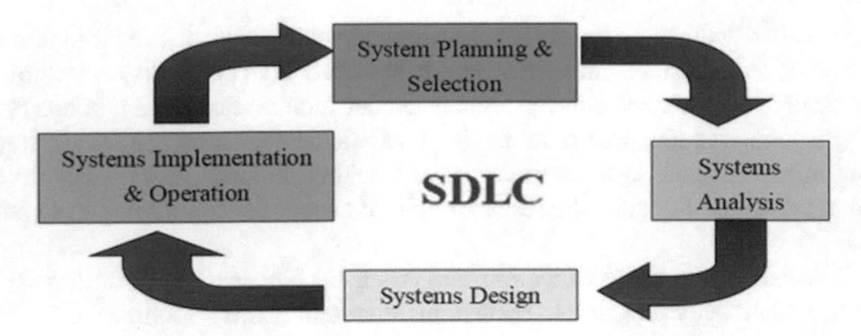

Fig. 1. Life cycle of the student achievement prediction system

2.2 System Use Case Diagram

Use case diagrams are an integral part of the system development life cycle, including actor and use-case symbols and connectors. Use-case models describe what a system does rather than on how, and reflect the system's view from a user's perspective outside of the system. The use case diagram of the student achievement prediction system is shown in Fig. 2.

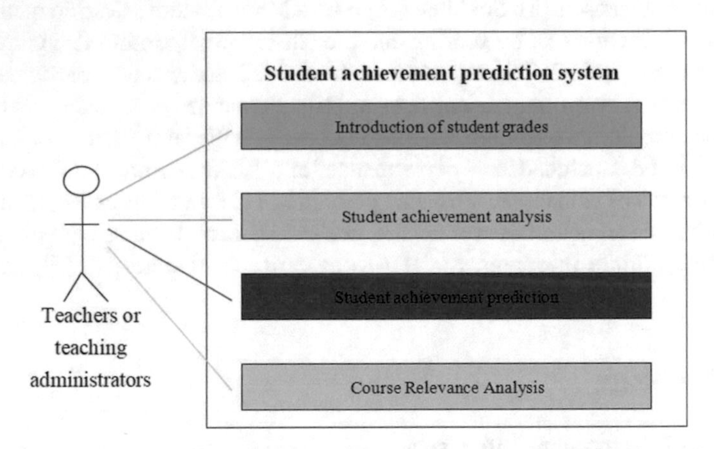

Fig. 2. Use-case diagram of the student achievement prediction system

2.3 Software and Hardware Requirements

The hardware requirement of the student achievement prediction system is a computer with an installed Windows operating system. The required software is Kettle software, SPSS Modeler software, and Python software. Specific hardware and software requirements are shown in Table 1.

Table 1. System hardware and software requirements

NO	Software or Hardware	Requirements
1	Computer	CPU: Processors with four or more cores are recommended Memory: 4 GB or more is recommended Hard Disk Drive: 500 GB or above Monitor: 24″ LED/LCD above recommended
2	Kettle software	Kettle 8.2.0 or above
3	SPSS Modeler software	IBM SPSS Modeler 18.0
4	Python software	Python3.8 or above

2.4 Data Analysis

The data analysis helps the researcher interpret various data sources, such as interviews and document reviews.

The statistical tools used are frequency and percentage for the study participants and the weighted mean for the ISO survey on system compliance. The researcher uses the 5-point Likert scale for the respondents' rating of the proposed system. Each item in the questionnaire is rated from 1 to 5 (5 is the highest and 1 is the lowest). The level of compliance is based on the ISO 25010:2011 quality software standard instrument of the proposed system. The scaling criteria shown in Table 2 are used.

Table 2. Scale of interpretation

Scale range	Verbal interpretation
4.20–5.00	Very Great Extent (VGE)
3.40–4.19	Great Extent (GE)
2.60–3.39	Moderately Extent (ME)
1.80–2.59	Little Extent (LE)
1.00–1.79	Very Low Extent (VLE)

3 Results and Discussion

3.1 User Login Interface

The user Login interface is shown in Fig. 3. The user enters their username and password, and then clicks the "Login" button to enter the corresponding interface.

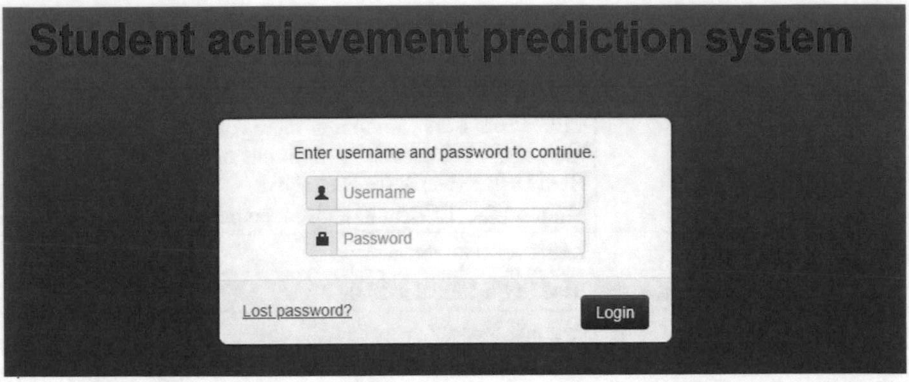

Fig. 3. User login module

3.2 Student Basic Information Administrator

The person in charge of student primary information data enters the username and password to enter the user's direct information input data operation interface (as shown in Fig. 4). The "Add Data" button and the data input dialog box pop up. After the user enters the information, they click the "Submit" button to submit the data.

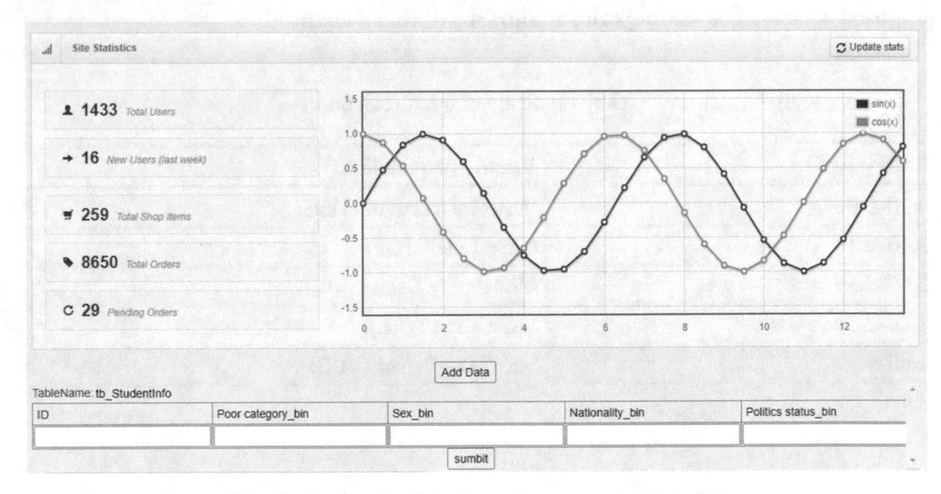

Fig. 4. Student basic information operation interface

Next, the user can click the "Select DataBase Table" button to enter the interface for viewing primary information data about the students (as shown in Fig. 5). Then, they can click the "Select" button to view basic information of all of the students, click the "Update" button to update data, and click the "Delete" button to delete data. They can then enter the student ID number in the "Search" editing box to query the basic information of a single student (as shown in Fig. 6).

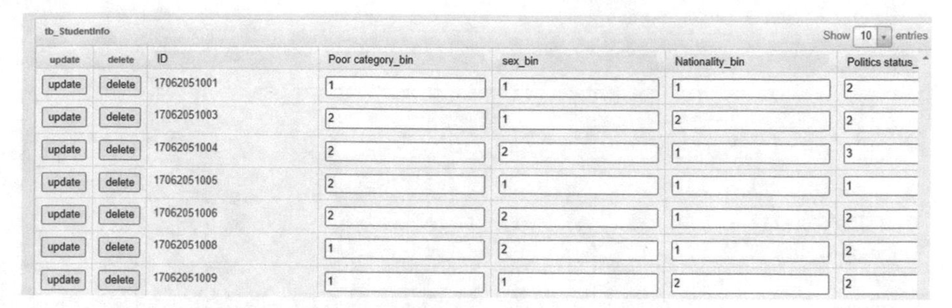

tb_StudentInfo — Show 10 entries

update	delete	ID	Poor category_bin	sex_bin	Nationality_bin	Politics status_
update	delete	17062051001	1	1	1	2
update	delete	17062051003	2	1	2	2
update	delete	17062051004	2	2	1	3
update	delete	17062051005	2	1	1	1
update	delete	17062051006	2	2	1	2
update	delete	17062051008	1	2	1	2
update	delete	17062051009	1	1	2	2

Fig. 5. Screenshot of the students' data information

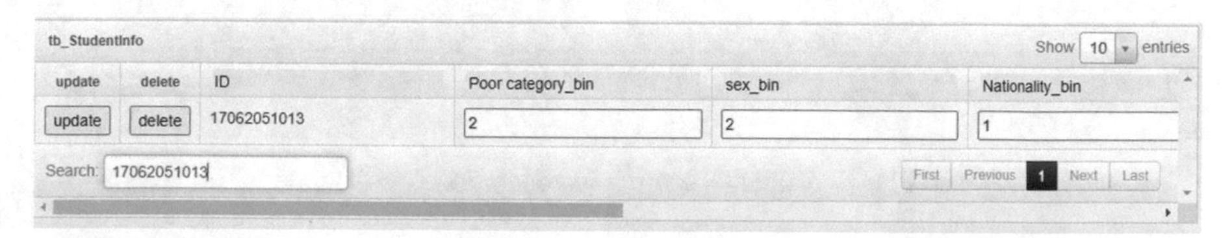

tb_StudentInfo — Show 10 entries

update	delete	ID	Poor category_bin	sex_bin	Nationality_bin
update	delete	17062051013	2	2	1

Search: 17062051013 — First Previous 1 Next Last

Fig. 6. Screenshot of a student's personal information with an ID number 17062051013

3.3 Student Course Score Manager

The user in charge of the student score input enters the username and password to log into the student course score input interface (as shown in Fig. 7.). They can click the "Add Data" button to enter the student's score input interface and click the "Submit" button to submit data after the score input.

Then, the user can click the "Select Database Table" button to enter the menu bar's student course score data table. They click the "Select" button to view the score information of all of the students (as shown in Fig. 8.). Then, they enter the ID number in the search box to view the data information of a single student (as shown in Fig. 9.). The user can click "Update" to modify data and click "Delete" to delete data.

3.4 Data Analysis Users

Cluster Analysis of Students. The user can enter the student clustering interface and select variables for student clustering analysis:

- Poor Category bin
- Sex bin,
- Nationality bin,
- Politics status bin,
- Whether to File card bin
- Urban and rural Students bin
- Library punch card records bin
- Age bin
- Living expenses bin

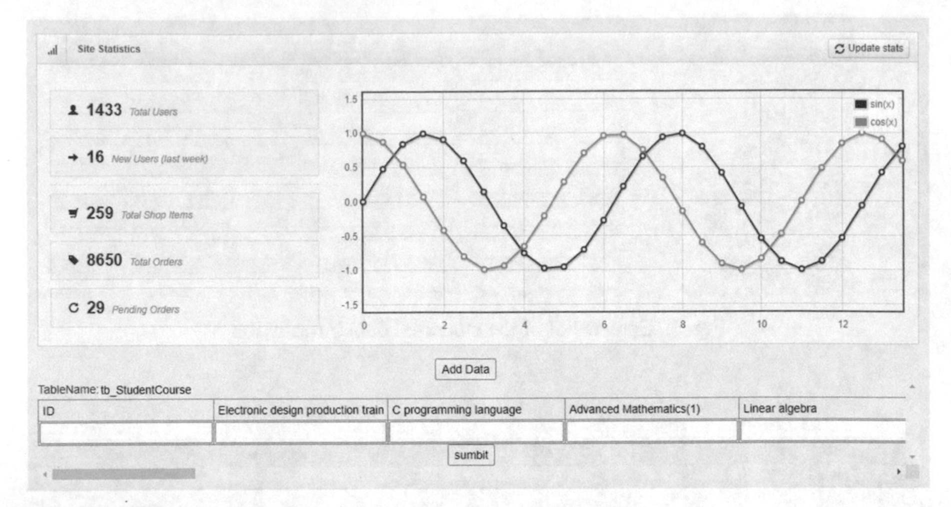

Fig. 7. Student's score input operation interface

tb_StudentCourse — Show 10 entries

update	delete	ID	Electronic design production train	C programming language	Advanced Mathematics(1)
update	delete	17062051001	81	83	69
update	delete	17062051003	84	73	69
update	delete	17062051004	85	89	86
update	delete	17062051005	85	76	85
update	delete	17062051006	84	83	81
update	delete	17062051008	84	73	72
update	delete	17062051009	83	60	61

Fig. 8. Information on all of the students' grades

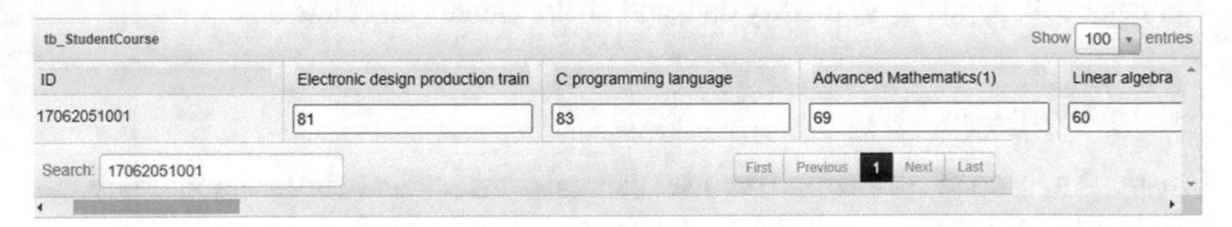
tb_StudentCourse — Show 100 entries

ID	Electronic design production train	C programming language	Advanced Mathematics(1)	Linear algebra
17062051001	81	83	69	60

Search: 17062051001 — First Previous 1 Next Last

Fig. 9. Screenshot of a student's score information with the ID no. 17062051001

- Book borrowing information bin

The selection results of the students' basic information fields are shown in Fig. 10. The K-means algorithm is used to conduct clustering analysis on the students. To ensure that the sample number of each type of student should not differ too much during the cluster analysis of students, the "random state" of the k-means is set to "35" after several rounds of debugging. The "OK" button is clicked to obtain the result of the student cluster analysis, as shown in Fig. 11.

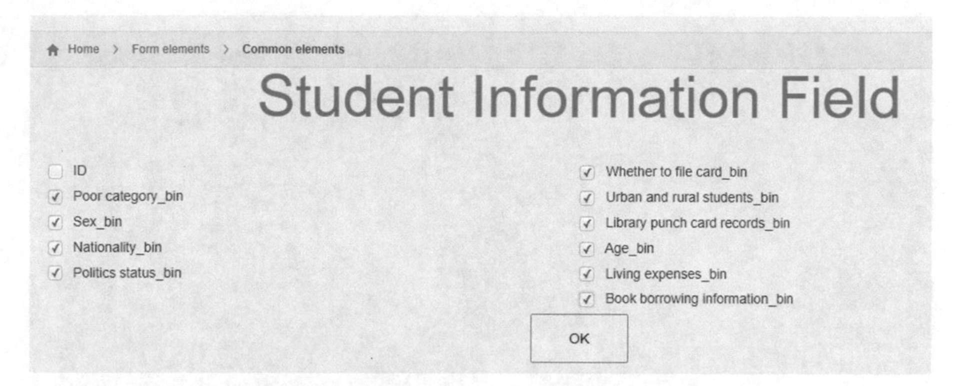

Fig. 10. Student basic information field selection

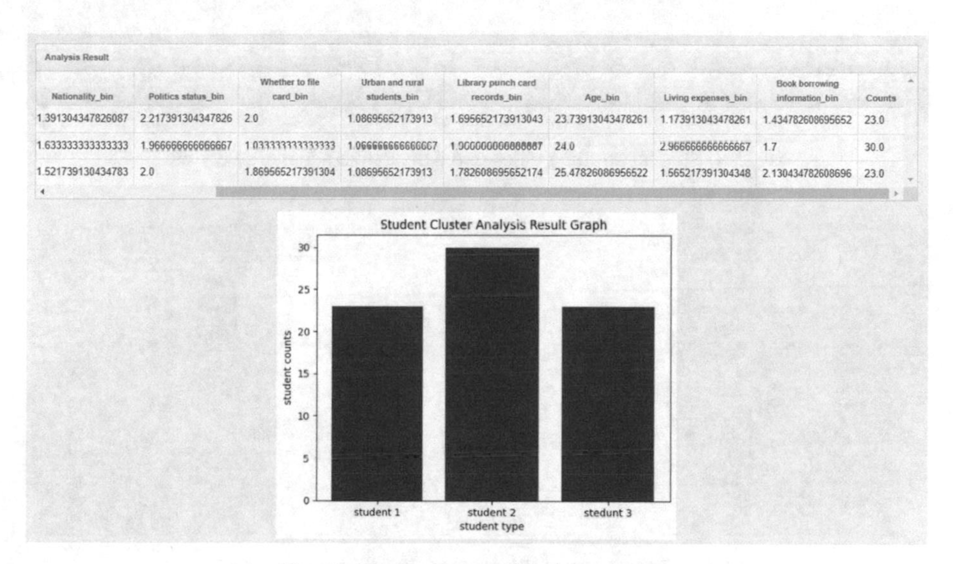

Analysis Result

Nationality_bin	Politics status_bin	Whether to file card_bin	Urban and rural students_bin	Library punch card records_bin	Age_bin	Living expenses_bin	Book borrowing information_bin	Counts
1.391304347826087	2.217391304347826	2.0	1.08695652173913	1.695652173913043	23.73913043478261	1.173913043478261	1.434782608695652	23.0
1.633333333333333	1.966666666666667	1.033333333333333	1.066666666666667	1.900000000000007	24.0	2.966666666666667	1.7	30.0
1.521739130434783	2.0	1.869565217391304	1.08695652173913	1.782608695652174	25.47826086956522	1.565217391304348	2.130434782608696	23.0

Fig. 11. Clustering results of the students

Discrete Treatment of the Students' Course Scores. The user can click on the "processing" course grade students discretization processing into the interface, and select requires a discretization processing course name (as shown in Fig. 12). According to each student's course grade using points box technology, student performance information for each course can be divided into three phases, where the course-specific scores (continuous data types) can be converted into the students' performance levels (categorical data types). On the student performance scale, "1" is defined as average, "2" is defined as good, and "3" is defined as excellent. The results of the course discretization are shown in Fig. 13.

The purpose of the student cluster analysis and the course discrete processing is mainly a procedure of data preprocessing for the subsequent performance prediction analysis using the decision tree and the SVM algorithm.

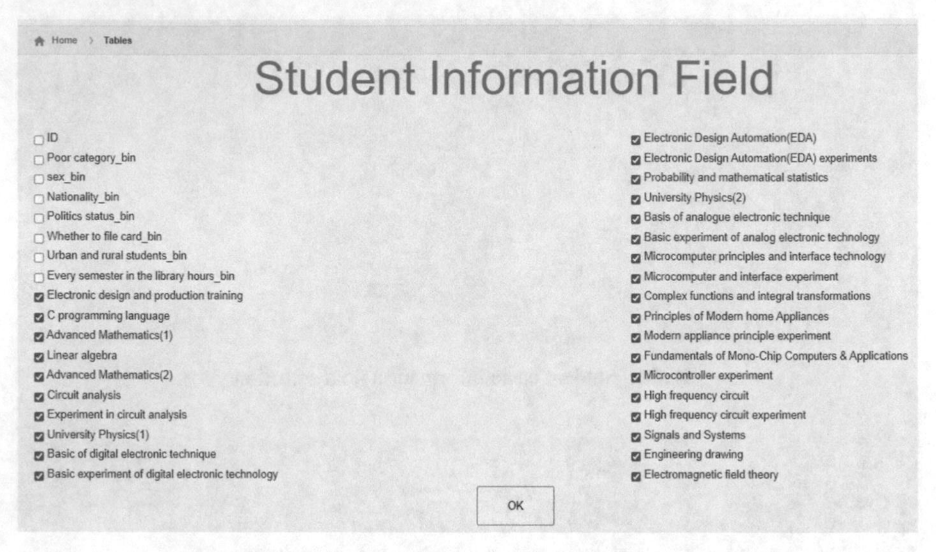

Fig. 12. Student course grades and a discretized course selection

Analysis Result — Show 10 entries

ID	Electronic design production training_BIN	C programming language_BIN	Advanced Mathematics(1)_BIN	Linear algebra_BIN	Advanced Mathematics(2)_BIN	Circuit analysis_BIN	Experiment circuit analysis_BIN	University Physics(1)_BIN	Basic digit electr techniqu
17062051001.0	1.0	3.0	1.0	1.0	3.0	2.0	2.0	2.0	1.0
17062051003.0	2.0	2.0	1.0	1.0	2.0	1.0	2.0	1.0	1.0
17062051004.0	2.0	3.0	3.0	1.0	3.0	1.0	3.0	2.0	3.0
17062051005.0	2.0	2.0	3.0	2.0	3.0	2.0	2.0	2.0	3.0
17062051006.0	2.0	3.0	3.0	2.0	3.0	1.0	3.0	2.0	2.0
17062051008.0	2.0	2.0	2.0	1.0	2.0	1.0	3.0	1.0	2.0

Fig. 13. Screenshot of discrete results of the course grades

Student Performance Prediction. In the prediction process of student performance, according to the types of students and the version of the pre-study courses, the decision tree and the SVM algorithms are used to find the main factors that affect students' performance in a particular period. Students' performance in the Basis of Analog Electronic Technique course was used as an example to conduct system debugging. To identify the important factors influencing the students' learning performance in "Fundamentals of Analog Electronic Technology," the selected predictive input variables are as follows:

- Electronic Design and Production training BIN

- C Programming language BIN
- Advanced Mathematics(1) BIN
- Linear in the input field Algebra BIN
- Advanced Mathematics(2) BIN
- Circuit analysis BIN
- Experiment in Circuit analysis BIN
- University Physics(1) BIN
- Basic of Digital Electronic technique BIN
- Basic Experiment of Digital Electronic Technology BIN
- Student Type BIN

The predicted output variable is:

- Basis of analog Electronic technique BIN

The selection of the input variable fields and the output variable fields is shown in Fig. 14. The user can click the "OK" button, and the performance prediction result of the Basis of Analog Electronic technique BIN is shown in Fig. 15.

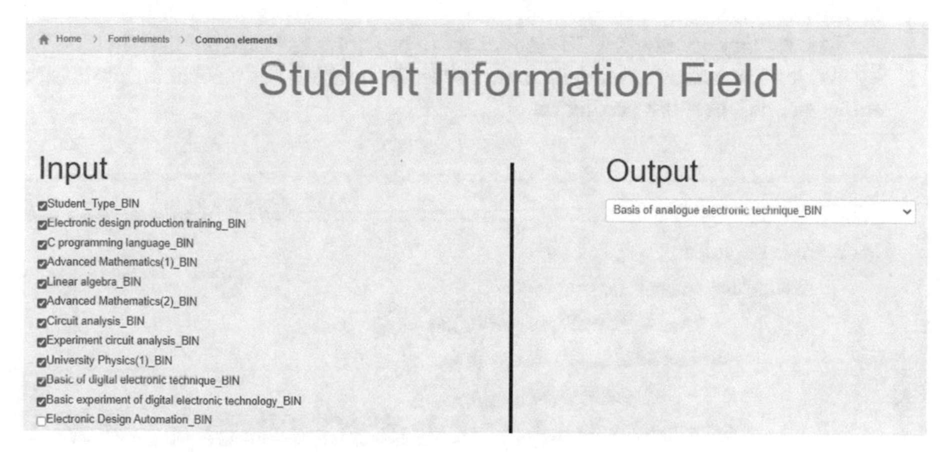

Fig. 14. Grade prediction input field and target field selection

These are the conclusions that can be drawn from the predicted analysis results:

(1) Predictive sensitivity analysis: The predicted sensitivity analysis figure is shown in Fig. 15. Among the 11 input variables, higher mathematics (2) BIN has the most substantial influence on students' analog electronic technology foundation, with an importance greater than 0.4. The second is the University Physics (1) BIN, with an importance of approximately 0.15. This is followed by Circuit Analysis BIN and C Programming Language BIN, with importance between 0.1 and 0.15. Therefore, in the Fundamentals of Analog Electronic Technology course, the main factors affecting the students' learning performance are as follows:

- Advanced Mathematics (2)
- University Physics(1) BIN
- Circuit Analysis BIN
- C Programming language BIN

(2) Prediction accuracy: During the model training process, the system's accuracy is 71.7%. During the model testing process, the system's accuracy is 78.3%.

(3) According to the prediction report (as shown in Fig. 16), the average performance of the eight samples is "1," the prediction accuracy is 89%, the recall rate is 100%, and the F1 score is 94%. The number of samples with good scores ("2") is 8, the prediction accuracy was 64%, the recall rate was 88%, and the F1 score was 74%. The number of excellent samples ("3") is 7, the prediction accuracy is 100%, the recall rate is 43%, and the F1 score is 60%.

(4) From the perspective of the confusion matrix (as shown in Fig. 17), there are 23 samples in the test, and 18 samples are correctly predicted, with an accuracy rate of 78.3%. Among them, 9 items with an average score of "1" participated in the test, 8 are correct, and the accuracy rate is 88.9%. There are 11 samples with good scores ("2"), 7 pieces are correctly predicted, and the prediction accuracy is 63.6%. Of the 3 samples with excellent scores ("3 "), 3 are correct, with an accuracy rate of 100%.

In summary, Python's KCS student score prediction system has certain reliability and high accuracy in student score prediction, which has a specific application value for student score prediction.

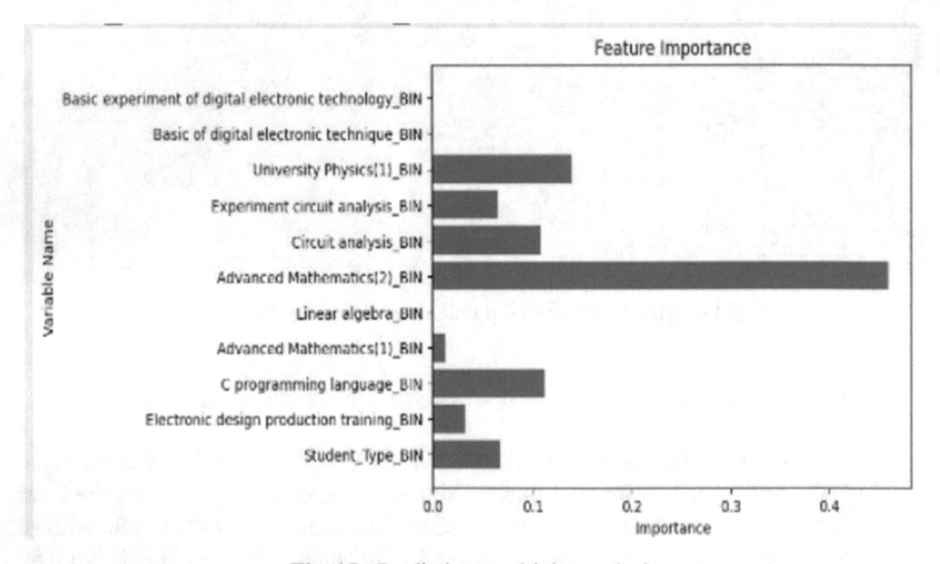

Fig. 15. Predictive sensitivity analysis

	precision	recall	f1-score	support
1	0.89	1.00	0.94	8
2	0.64	0.88	0.74	8
3	1.00	0.43	0.60	7
accuracy			0.78	23
macroavg	0.84	0.77	0.76	23
weightedavg	0.83	0.78	0.77	23

score_train: 0.717 score_test: 0.783

Fig. 16. Forecast accuracy and forecast analysis report

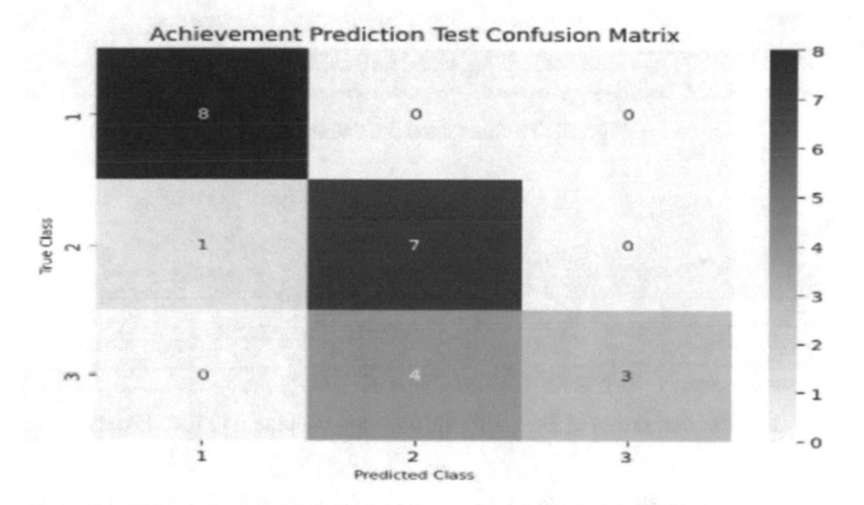

Fig. 17. Confusion matrix diagram of the predictive analysis

Check Student Data Information. The data analyst can view the data table shown in Fig. 18, where tb StudentInfo is the basic information table and tb StudentCourse is the score table. Tb StudentDataInfo is the statistics table with the student information (including primary student data and course score data). The data analyst can view the data information of any data table, as shown in Fig. 19, which is a screenshot of the statistical data table of student information.

The student ID number is entered into the search box to view the specific data information of the student, as shown in Fig. 20.

tb_StudentCourse	select
tb_StudentDataInfo	select
tb_StudentInfo	select

Fig. 18. Screenshot of the system data table

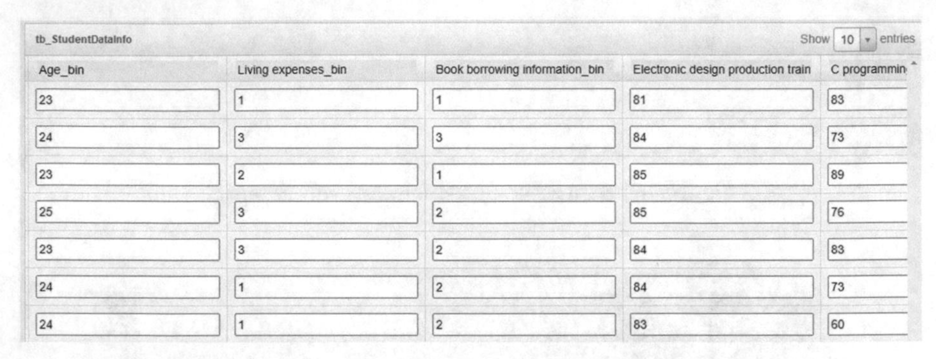

tb_StudentDataInfo — Show 10 entries

Age_bin	Living expenses_bin	Book borrowing information_bin	Electronic design production train	C programmin
23	1	1	81	83
24	3	3	84	73
23	2	1	85	89
25	3	2	85	76
23	3	2	84	83
24	1	2	84	73
24	1	2	83	60

Fig. 19. Tb student datainfo screenshots

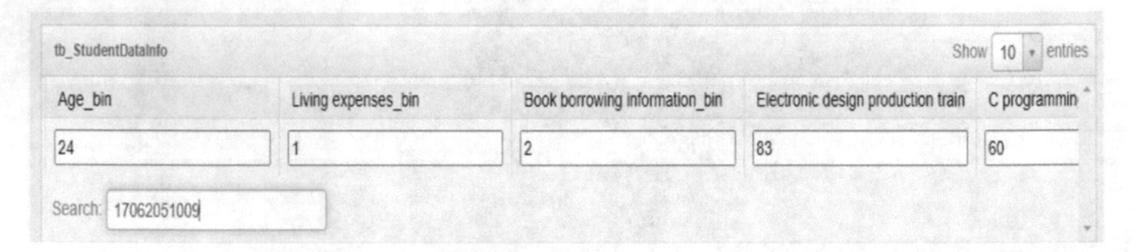

tb_StudentDataInfo — Show 10 entries

Age_bin	Living expenses_bin	Book borrowing information_bin	Electronic design production train	C programmin
24	1	2	83	60

Search: 17062051009

Fig. 20. Screenshot of the student information with the ID 17062051009

3.5 Extent of Compliance of the Developed System to the ISO 25010 Software Quality Standards

Table 3 presents the statistical table of the Student Achievement Prediction System's compliance with the ISO/IEC 25010 standards. As seen in the table, the results show that the eight (8) ISO/IEC 25010 Criteria of the developed Student Achievement Prediction System have means ranging from 4.45 to 4.62, which are all characterized as "Very Great Extent." Of the eight (8) ISO/IEC 25010 criteria, the attribute, "Compatibility," has the highest category mean of 4.62, which is descriptively interpreted as "Very Great Extent." However, the attribute, "Usability," has the lowest category mean of 4.45, which is also descriptively interpreted as "Very Great Extent." The overall mean of 4.54 suggests that the developed system is generally characterized as "Very Great Extent" in compliance with the ISO/IEC 25010.

Table 3. Summary of evaluation of the system's compliance with the ISO/IEC 25010 criteria

ISO/IEC 25010 criteria	Mean	Descriptive rating
Functional suitability	4.57	VGE
Performance efficiency	4.57	VGE
Compatibility	4.45	VGE
Usability	4.62	VGE
Reliability	4.50	VGE
Security	4.60	VGE
Maintainability	4.50	VGE
Portability	4.50	VGE
Overall mean	4.54	VGE

4 Conclusion

This study combines the k-means clustering algorithm, decision tree, and the SVM algorithm to develop the KCS student achievement prediction system and the algorithm model. The K-means clustering algorithm mainly divides students into three types according to their basic information, such as gender, ethnicity, source of urban and rural students, and poverty. The SVM algorithm is mainly used to improve the accuracy of the student achievement prediction system. Additionally, the decision tree algorithm is used mainly for the sensitivity analysis of student score prediction to help educators find the main factors that affect students' grades.

The student achievement prediction system based on the KCS algorithm model can provide decision-making guidance and suggestions for educators and educational administrators engaged in electronic information engineering education in future education management. Educators and education administrators can use the system to predict the situation of the follow-up courses according to the types of students and techniques that they have employed. According to the importance of the predictive variables in the decision tree, the learning content and teaching methods should be carried out for an appropriate review of students who have not achieved ideal results in the advanced courses.

References

1. Ding, G.Y.: Research on the Data Modeling of Academic Performance of College Students - Based on the Analysis of College Educational Data. Nanjing Normal University, China (2019)
2. Okike, E.U., Mogorosi, M.: Educational data mining for monitoring and improving academic performance at university levels. Int. J. Adv. Comput. Sci. Appl. **11**(11), 570–581 (2020)
3. Robbi, R.: Educational Data Mining (EDM) on the use of the internet in the world of Indonesian education. TEM J. **9**(3), 1134–1140 (2020)
4. Tang, H.T., Xing, W.L., Pei, B.: Time really matters: understanding the temporal dimension of online learning using educational data mining. J. Educ. Comput. Res. **57**(5), 1326–1347 (2019)

5. Tong, J.F.: Research on accurate early warning mechanism of online learning based on educational data mining. Comput. Prod. Circ. (05), 231 (2019)
6. Lv, H.Y., Zhou, H.J., Zhang, J.: Research on the application of educational data mining in the analysis of students' online learning behavior under the background of big data. Comput. Technol. Autom. **36**(01), 136–140 (2017)
7. Shi, Q., Qian, Y., Sun, L.: Research on online learning process supervision based on educational data mining. Mod. Educ. Technol. **26**(06), 87–93 (2016)
8. Zaffar, M., Hashmani, M.A., Savita, K.S., Rizvi, S.S.H., Rehman, M.: Role of FCBF feature selection in educational data mining. Mehran Univ. Res. J. Eng. Technol. **39**(4), 772–778 (2020)
9. Sokkhey, P., Okazaki, T.: Developing web-based support systems for predicting poor-performing students using educational data mining techniques. Studies **11**(7) (2020)
10. Güre, Ö.B., Kayri, M., Erdoğan, F.: Analysis of factors effecting PISA 2015 mathematics literacy via educational data mining. Egitim ve Bilim **45**(202) (2020)
11. Fernandes, E., Holanda, M., Victorino, M., Borges, V., Carvalho, R., Van Erven, G.: Educational data mining: Predictive analysis of academic performance of public school students in the capital of Brazil. J. Bus. Res. **94**, 335–343 (2019)
12. Fan, L.: A student achievement prediction model based on big data under the background of intelligent education. Inf. Comput. (Theory Ed.) **31**(24), 223–225 (2019)
13. Zhou, Q., Wang, W.F., Ge, L., Xiao, Y.F., Tang, D.: Student achievement prediction based on one-card data and course classification. Comput. Knowl. Technol. **14**(24), 236–239 (2018)
14. Song, Y.T.: A student achievement prediction model based on one-card data and multiple linear regression. Pract. Electron. (24), 57–58+85 (2017)
15. Sun, Q.L.: The Research of Network Log for Student Achievement Prediction. Chongqing University (2017)
16. Zhen, Y.J.: Research and Realization of College Student Achievement Prediction System Based on Network Logs. Chongqing University (2016)
17. Kendall, K.E., Kendall, J.E.: Systems Analysis and Design. Pearson Prentice Hall, Upper Saddle River (2011)

Prediction and Analysis of COVID-19's Prevention and Control Based on AnyLogic in the Background of Big Data

Cuoling Zhang(✉) and Deqing Zhang

Anhui Sanlian University, Hefei, China
zhangc@students.national-u.edu.ph

Abstract. Based on the surging situation of COVID-19 and its rapid propagation speed and wide range, schools as a crowded place are prone to outbreak of large areas. Therefore, in order to ensure the normal progress of campus teaching order, campus should be the focus of epidemic prevention and control. By using AnyLogic simulation software, using system dynamics model and combined with the actual data operation of College D Teaching Building, this paper simulates the D Teaching Building, and intuitively shows the simulation and control effect through the Time Plot. The simulation results show that the school can reduce the number of infected people by taking effective measures to control the contact rate of students and vaccinating them in time; Finally, effective treatment can greatly increase the rehabilitation rate of infected people and reduce the number of deaths. This method has certain guiding significance in today's severe epidemic prevention and control.

Keywords: COVID-19 · System dynamics · SEIR model · Big data analysis

1 Introduction

COVID-19 originated in Wuhan, Hubei Province, China in December 2019. With the flow of people, the virus spread rapidly and quickly spread to all countries around the world. The spread of the virus is difficult to control and has a serious impact on economy, transportation, work, life and many other aspects [1]. At present, the spread of the epidemic has not been fully controlled. The stronger infectivity of the Omicron variant has led to a surge in the number of new coronavirus infections worldwide. According to the results of the implementation of big data report on the epidemic, the number of confirmed cases worldwide on April 23, 2022 was as high as about 650000, with a total of more than 500000 confirmed cases. According to the data report of the global epidemic prediction system of Lanzhou University, this round of epidemic will last until 2023 [2]. According to the statistics of global network, there are more than 40 million college students in China [3]. College students are the pillars of the country and the main force of national defense technology in the future. Moreover, the school has the characteristics of closed space and dense personnel. Therefore, the school is the key place for epidemic prevention and control.

© The Author(s), under exclusive license to Springer Nature Switzerland AG 2023
L. C. Tang and H. Wang (Eds.): BDET 2022, LNDECT 150, pp. 75–84, 2023.
https://doi.org/10.1007/978-3-031-17548-0_7

2 Significance of Problem

Although it has been three years since COVID-19 was first discovered, the epidemic has not been completely controlled, and the virus has mutated and fought back. Recently, there have even been confirmed cases among teachers and students in more than one school. As the most densely populated area and the most prone to cross-infection after the resumption of classes due to the epidemic, public teaching buildings in colleges and universities will not only affect the physical and mental health and academic performance of students, but even affect the teaching order of the school. Teaching buildings usually do not have fixed main entrances or exits or are crowded. Therefore, teaching buildings will be the focus and difficulty of epidemic control. The number of student groups is large and the mobility is large. There are usually no fixed main entrances and exits in public places on campus, or crowded crowds at the entrances and exits will be difficult to control. The traditional management method is based on experience judgment and case reference. It is difficult to comprehensively predict the flow of students in the student group, and the effect of its control measures is often difficult to predict. Once an infection case occurs, the entire student group will be difficult to control, and the consequences will be unimaginable.

AnyLogic is a modeling tool that supports modeling based on system dynamics, and can combine three models for modeling, supports a visual development environment, supports drag-and-drop operations of modules, and supports Java programming. Therefore, using AnyLogic to simulate the epidemic situation of the Teaching Building D in our school will help students to understand the spread of the virus, enhance the awareness of the epidemic prevention and control of the student group and other personnel, reduce the probability of epidemic spread, and reduce the disease as much as possible. The threat and harm to students has important theoretical and research significance.

3 Literature Review

The process of modeling based on system dynamics is introduced in detail in "anonymous in 3 days" [4]. The research of this project is to take this model as a reference, combine the characteristics of the teaching building as the research object, conduct demand analysis, simulate the system dynamics model in days, use the AnyLogic platform, combine the real data of novel coronavirus pneumonia, carry out simulation research on epidemic prevention and control, and evaluate its development trend.

Through the query of network data, we know that the infection rate of COVID-19 is about 0.59, which is the basis for setting up the model parameters of [5].

Guoqiang Wu and Limin Jiang's article on "Epidemic Prevention and Control Manual of Campus Building and Environment": The overall prevention and control scheme, prevention and control zoning, protection standards, organization and management requirements of the campus given in this paper put forward specific prevention and control principles and measures for campus teaching area, living area and outdoor places, which can be used as a reference for our system simulation [6].

Through the study of bilibili network video materials, I learned the pedestrian simulation of subway station and understood the powerful function of AnyLogic pedestrian library [7].

Qiaoming Deng and Chengzhi Lan and Yubo Liu's article on "Simulation study on epidemic Control in public Teaching Building of University Campus based on AnyLogic platform"– A case study of Building 34 of South China University of Technology: This paper clearly describes the first mock exam of the epidemic control and management of university teaching building using AnyLogic platform. The simulation system uses the hybrid modeling method based on discrete events and agents and gives the statistical chart of the flow rate and the infection line map, which provides a reference for our research [8].

4 Modelling Questions

Due to the phenomenon that traditional management and control relies on experience, it is difficult to monitor comprehensively, and it is difficult to predict the efficiency of management and control. Therefore, the research of this project intends to solve the following problems:

(1) Once a student is infected, the relationship between the number of contacts between students and the peak infection.
(2) From the perspective of epidemic modeling, the impact of contact rate on mortality and the change of infection rate on mortality before and after the adoption of prevention and control measures are analyzed.

5 Modeling Methodology

Taking the D Teaching Building of Anhui SanLian College as an example, this project uses the AnyLogic 8.7.10 environmental simulation modeling platform to simulate the epidemic spread of D Teaching Building. D Teaching Building is located in the southwest corner of the campus and connected to the student accommodation area through the pedestrian overpass. D Teaching Building is connected with E and H teaching buildings. There are multiple entrances and exits. During class and after class, the personnel mobility and contact rate are high, which is more complex for epidemic control. This study simulates and predicts the situation of students in D Teaching Building, and explores the impact of prevention and control measures on the spread of the epidemic in the teaching building; At the same time, the Time Plot and sensitivity analysis chart are also used to intuitively show the effect of control measures.

Modelling approach: Modeling methods using system dynamics
Simulation time units: days (1 second's simulation time = 1 days of actual time)
Duration of simulation: 100 days.

5.1 Data Information Collection

By consulting the timetable of the Teaching Building and visiting the student dormitory, it is concluded that the average daily number of students in Teaching Building D is about 1000, and the number of people who may have close contact with the dormitory is 1–10. According to the survey of network big data, the possible infection probability of each college student is about 0.3 (the infection rate in the United States is 0.58 [9]), the incubation period of the virus is generally about 14 days, the treatment time is about 15–30 days, and the mortality is about 0.5%–1% [10].

5.2 Determine the Required Stock and Establish the Model

Initialize the Model. The SEIR model divides the population in the epidemic area of infectious diseases into four categories: Susceptible, Exposed, Infectious, and Recovered. Based on the model of infectious diseases, the research and promotion of infectious disease dynamics models are generated. Quantitative analysis and numerical simulation are used to analyze the development process of the disease, predict the change trend, and analyze the reasons for the spread of the disease, which are consistent with the COVID-19 situation to be studied. The actual situation of the development, on which the death category stock is added.

First, the first step in modeling is to determine the stock: Susceptible, Exposed, Infectious, Recovered, Death. Then add the relevant parameters mentioned. In order to establish the relationship between the stocks, the following 4 flows are added: ExposedRate, InfectiousRate, RecoveredRate, and Mortality Rate are established. The calculation formula is as follows:

$$\frac{dS}{d\mathrm{t}} = -\alpha \frac{SI}{N} \tag{1}$$

$$\frac{dE}{dt} = \alpha \frac{SI}{N} \tag{2}$$

$$\frac{dI}{dt} = \beta E \tag{3}$$

$$\frac{dR}{dt} = \gamma I \tag{4}$$

$$\frac{dD}{dt} = \delta I \tag{5}$$

where N = S + E + I + R + D (S = N at the beginning of the virus), α Indicates S in days α The speed of changes to E, β Indicates E in daily β The speed of becomes I, γ Indicates I in daily γ The speed of changes to R, δ Means I every day in δ The speed of changes to D [11].

According to the collected daily average number of students in class is 1000, the initialization model is shown in Fig. 1.

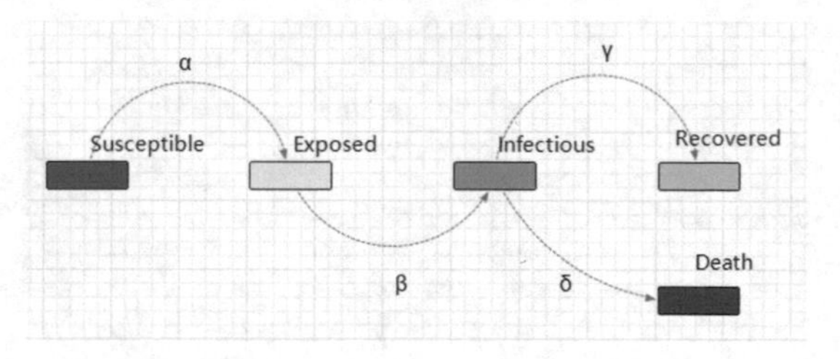

Fig. 1. Initializing the COVID-19 model

Modify the Model Parameters to Better Simulate the Operation of the Teaching Building. Taking into account the reduction of the disease transmission rate, some epidemic prevention and control measures are taken, mainly to control the average daily number of contacts of students. In addition, the immunity of the students themselves also has a positive effect on the cure of the disease. Adjust the parameter changes by increasing the slider, the parameter range of the slider is as follows [12]:

ContactRatcInfcctiou = 1–10 people
Infectivity = 0.1–0.6
AverageIncubationTime = 1–14 days
AverageIllnessDuration = 5–20 days
CureRate = 94.3%
Mortality = 0.3%–1.5%
As shown in Fig. 2.

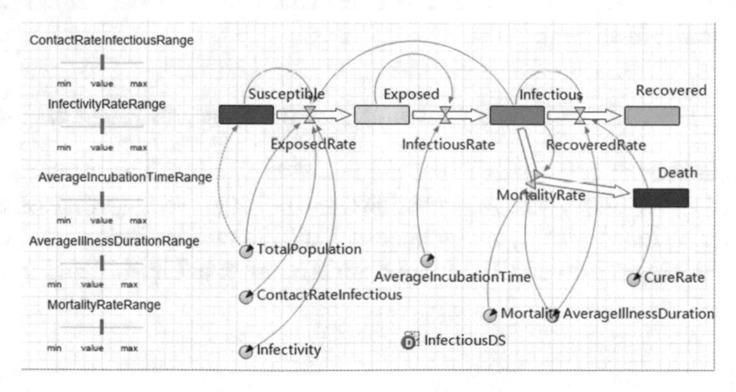

Fig. 2. Model parameter slider

Add a Time Plot chart to visually output the changes in the number of students in each state over time, as shown in Fig. 3.

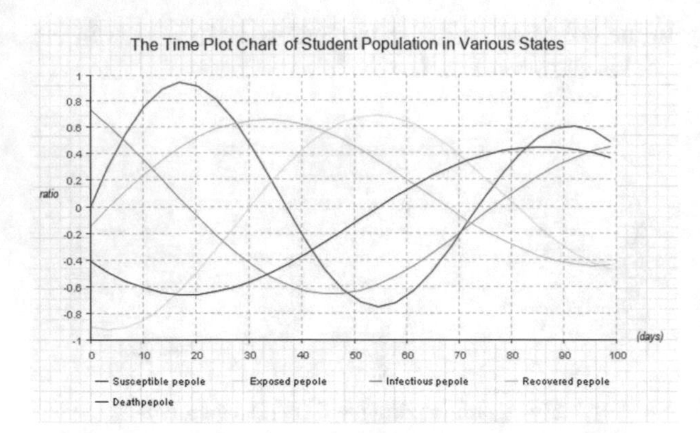

Fig. 3. Time plot chart

6 Results and Analysis

Through the system dynamics model and sensitivity analysis, it can be seen that if there are unfortunate infected people among the students, taking effective measures to control the contact rate of the students, such as controlling the students going out and reducing the contact between the students, can effectively control the epidemic situation and reduce the Infection of other students; through effective vaccination, the immunity of students can be enhanced, and the number of infected can also be reduced; finally, through effective treatment, the recovery rate of infected people can be greatly increased and the number of deaths can be reduced. The specific experimental results and data are as follows:

Special note:

The simulated facility is: D Teaching Building

The facility location of this research is: AnHui SanLian University, Hefei City, China

Total number of student: 1000.

6.1 The Number of Contacts is Constant, and the Infection Rate Changes

When the number of contacts is 3 and the infection rate is 0.3, the other parameters are assumed as follows: AverageIncubationTime = 7days, AverageIllnessDuration = 7days, MortalityRate = 0.5%, the number of deaths in the simulation results is 5.264. The model running results and Time Plot chart are shown in Fig. 4(a) and (b).

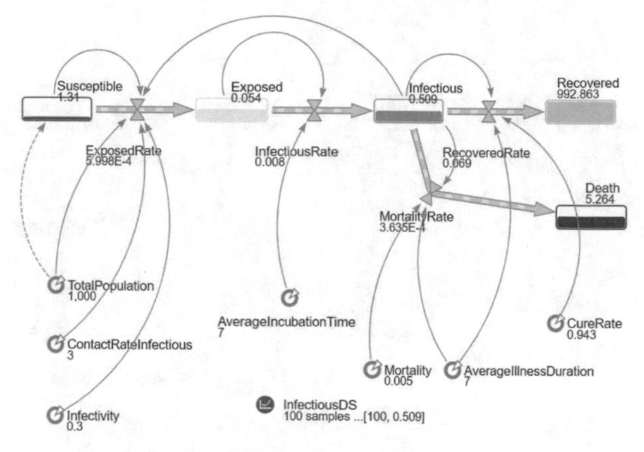

(a) Result Diagram of Model Operation

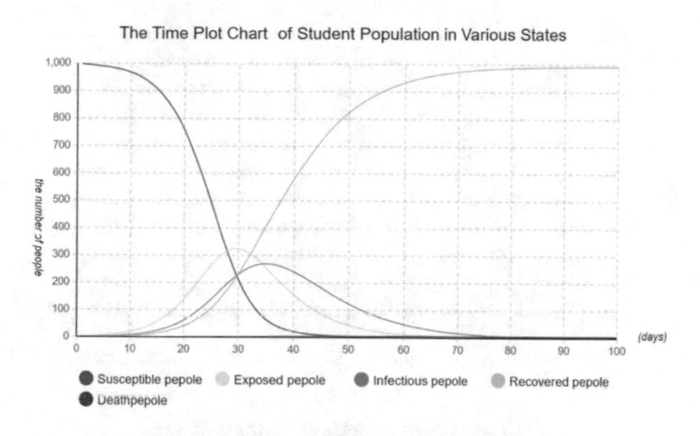

(b) Line Chart of Model Rnning Time

Fig. 4. Model running graph and time plot chart when the number of contacts is 3 and the infection rate is 0.3

Similarly, when there are 3 people in contact and the infection rate is 0.6, the other parameters are the same as above, and the number of deaths in the simulation results is 5.273. The model running results and line graph are shown in Fig. 5(a) and (b).

6.2 The InfectionRate is the Same, and the Number of Contacts Changes

Assuming that the InfectionRate is still 0.3, but the number of contacts is 10, the other parameters are the same as above, and the number of deaths in the simulation results is 5.274. The model running results and line graph are shown in Fig. 6(a) and (b).

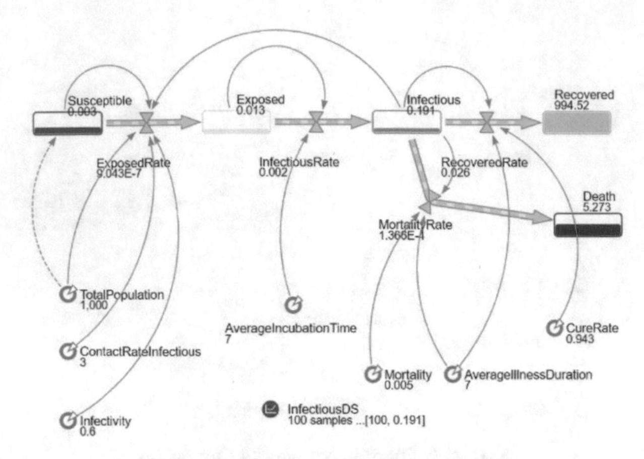

(a) Result Diagram of Model Operation

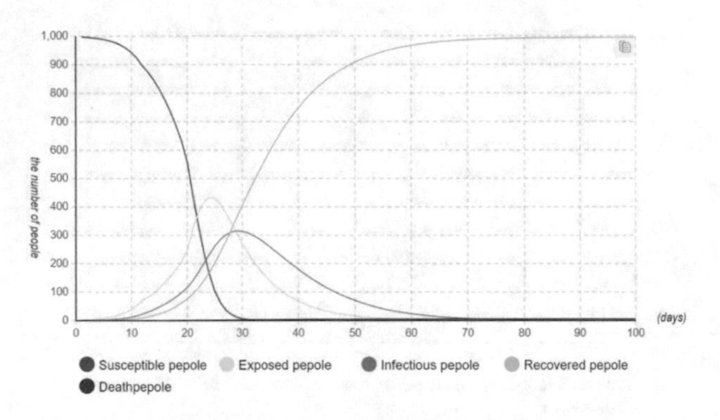

(b) Line Chart of Model Rnning Time

Fig. 5. Model running graph and time plot chart when the number of contacts is 3 and the infection rate is 0.6

6.3 It Can also Be Seen from the Sensitivity Analysis Chart

That when CR = 10(Note: CR indicates ContactRate), about 35 days, the number of infected people reached a peak of about 275; when CR = 4, about 73 days, the number of infections reached a peak of about 150 people. Therefore, effective control of the contact rate can slow down the spread of the epidemic and significantly reduce the number of infected people, as shown in Fig. 7. The abscissa represents the number of days and the ordinate represents the number of people.

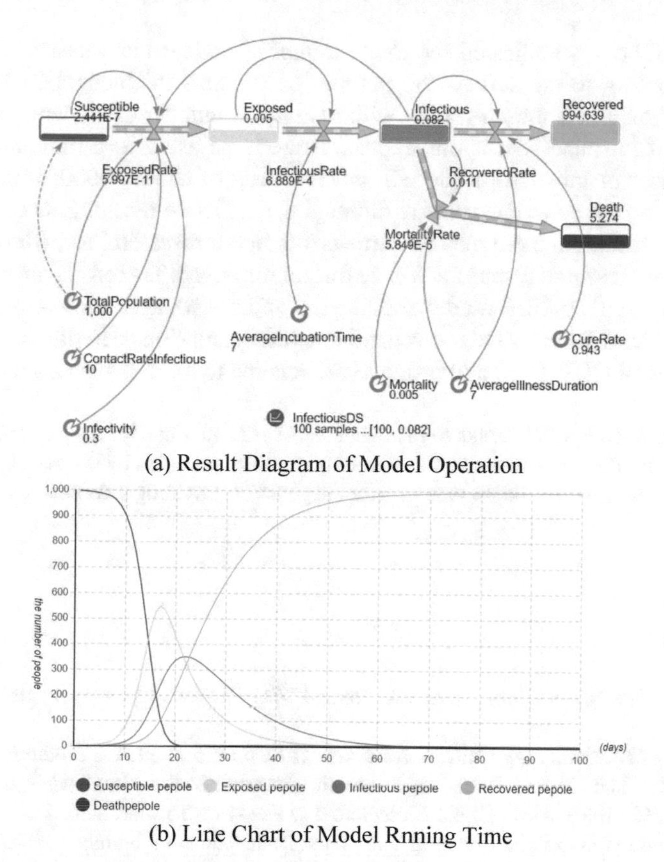

(a) Result Diagram of Model Operation

(b) Line Chart of Model Rnning Time

Fig. 6. Model running graph and time line graph when the number of contacts is 10 and the infection rate is 0.3

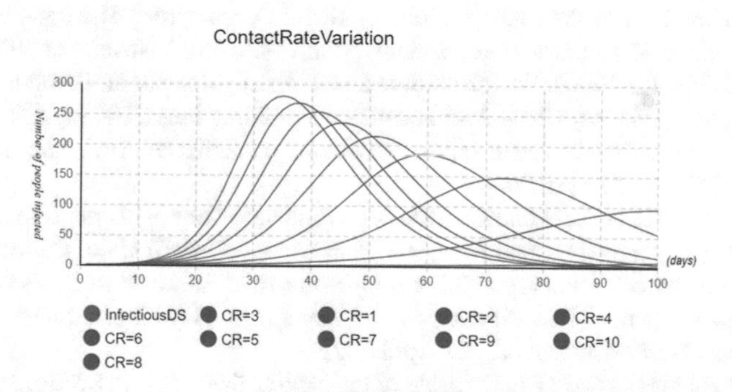

Fig. 7. Sensitivity analysis plot

7 Conclusion and Recommendation

Through experiments, it is proved that in today's severe epidemic prevention and control, the System Dynamics model plays a vital role in the simulation of epidemic prevention and control. Then, we adjusted the total number of colleges and universities to about 70000 (according to the survey, the number of students in Zhengzhou University is about 72600, which is the university with the largest number of students in China) for simulation. It is found that if our exposure rate is adjusted to a maximum of about 52, the number of infected people will reach a peak of about 20000; If students have positive results, the consequences are unimaginable. Due to the continuous variation of the epidemic situation, there may be some errors in our research data, which makes the accuracy of our research results need to be further improved. Therefore, our next research plan is to improve the SEIR model and find the optimal configuration of parameters for experimental simulation, so that our experimental conclusions can play a better role in the protection of COVID-19 and promote the scheme to more colleges and universities.

Acknowledgment. Special thanks to Professor Leorey O. Marquez for his guidance and help in the field of computer simulations and the learning platform provided by the National University. At the same time, I would like to thank my colleagues for providing reference materials when I do this topic.

References

1. Xiao, C.: The Novel Coronavirus was Named "COVID-19". Guangming Net, 13 February 2020
2. Institute of Microbiology Chinese Academy of Sciences: The latest Prediction of Lanzhou University: The Current Round of Epidemic Situation in the Mainland is Expected to be Controlled in Early April. China Biotechnology Network, 15 March 2022
3. World Wide Web Official Account: The Latest Data: The Total Number of College Students in China is More than 40 Millions. Global Network, 20 May 2020
4. Originality Document. Learn AnyLogic Chinese Version in Three Days.pdf. Knowledge Sharing Platform, pp. 147–164 (2021)
5. Hong, J.: How High is the Infection Rate of Novel Coronavirus? The Results are Obtained from the Data of 45 Countries and Regions. Xianji Network, 5 November 2020
6. Wu, G., Jiang, L.: Epidemic Prevention and Control Manual of Campus Building and Environment, pp. 43–47. China Architecture and Building Press, Beijing (2020)
7. Dongfeng incense breaks seven miles. Detailed explanation of AnyLogic Subway Station Modeling. Bilibili, 23 April 2020
8. Deng, Q., Lan, C., Liu, Y.: Simulation Study on Epidemic Control in Public Teaching Building of University Campus Based on AnyLogic Platform – A Case Study of Building 34 of South China University of Technology. China Architecture and Architecture Press (2020)
9. CCTV News Client. CDC: About 58% of Americans Have Been Infected With Novel Coronavirus. Northeast Network, 27 April 2022
10. Zhou, P.: What is the Case Fatality Rate of the New Crown? Health Sector, 19 July 2020
11. Xiaobai. Simple Prediction of Epidemic Spread Based on SEIR Differential Equation Model. CSDN, 20 April 2022
12. Xinhua News Agency: China's Action against the COVID-19. The Central People's Government of the People's Republic of China, 7 June 2020

Distributed Multi-source Service Data Stream Processing Technology and Application in Power Grid Dispatching System

Tongwen Wang[1], Hong Zhang[2(✉)], Jinhui Ma[1], and Xincun Shen[1]

[1] State Grid Anhui Electric Power Co., Ltd., Hefei, China
[2] Nanjing Branch, China Electric Power Research Institute Co., Ltd., Nanjing, China
zhanghong2@epri.sgcc.com.cn

Abstract. Smart grid communication system carries a large number of business data of power grid terminals. With the wide access of new energy power, after the power grid is connected to renewable energy power, the data services required by the dispatching, control, management and other services of microgrid system need to be processed through the distribution network communication system. How to dynamically, intelligently and adaptively adjust the distribution of data flow service transmission and make each node work orderly is a new requirement of intelligent distributed data processing. This paper expounds the latest technologies from two aspects of transmission and calculation in the process of data stream processing, mainly including: data stream real-time transmission technology, service stream real-time processing system architecture, channel multiplexing processing method based on stream density, service stream real-time processing based on container technology, and introduces the application of these technologies in power grid system.

Keywords: Service stream real-time processing · Distributed multi-source service data stream · Smart power grid · Data stream real-time transmission

1 Characteristics of Data Flow

With the development of data acquisition technology and data processing technology, a new data model appears. Different from the traditional data modeling method, this kind of data is suitable for transient data flow modeling. Applications for data flow include network data monitoring, log flow mining, wireless sensor network detection and so on. In the data flow model, individual data items appear in the form of tuples, and multiple fields of tuples represent multiple attribute values of data items. Stanford University defines "streaming data" as a continuous, uninterrupted and unstructured data queue [1, 2]. It is considered that stream computing is a new computing mode that uses parallelism and location efficiently, uses stream computing processor, stream computing programming language and other technical means to process ordered data. In the data flow model, the input data is not stored in disk or memory at first. They arrive in the form of "continuous data flow". The processing mode of data stream is different from the

© The Author(s), under exclusive license to Springer Nature Switzerland AG 2023
L. C. Tang and H. Wang (Eds.): BDET 2022, LNDECT 150, pp. 85–94, 2023.
https://doi.org/10.1007/978-3-031-17548-0_8

traditional data processing mode. The main differences are as follows: the data stream is generated in real time, flows into the system in real time, and the analysis results are required to be given in real time. The processing system cannot make any assumptions about the generation and arrival of data flow. For example, the system cannot predict the scale of data flow per unit time, nor can it control the data flow to arrive at the system in order. Data flow is generated and flowed in indefinitely and continuously. The data stream flows through the system and processed at one time, and then it is difficult to recover the data stream, that is, if there is no data persistence mechanism after the data item is processed, it will be discarded. Many operations in the data flow model are semantically similar to relational operations, such as finding, selecting, connecting and so on. According to most stream data application scenarios, this paper defines a single data item in the data stream as follows [3, 4]. The data item in the data stream can be regarded as a binary composed of discrete and orderly timestamps and representing the mode relationship of the data item, we can also convert data items into this form by processing the logic of data source generation or data cleaning after data generation. Because this representation method can most completely identify the characteristics of a data item. Multiple parts identify multiple characteristics of data to distinguish other tuples, and the real information of data items is reflected in. Timestamp attribute is an important aspect that distinguishes data flow mode from traditional data mode. It represents that data items have priority in time. In streaming computing system, the processing of data is passively based on the order of data arrival, and the calculation results cannot be optimized by changing the input order of data. In order to avoid this problem, we can only change the execution order of task operators in the system in a disguised form, which also promotes the in-depth study of task scheduling algorithm in distributed system.

2 Overview of Flow Computing

Data flow calculation comes from a belief that the value of data decreases with the passage of time, so it must be processed immediately after the event occurs, rather than cached and processed in batches. In the traditional data processing mode, the data is often independent of the application. The system is responsible for storing the data in the relational table on the disk. The stored data is a static and fixed set. Although the application of this processing mode may query frequently, because the operation object is persistent and stable, the system can obtain all the data at a certain query time point. However, the mature technology of traditional distributed data processing is not suitable for the flow computing mode. The core value of flow computing is to integrate the data from a variety of heterogeneous data sources in real time and process the massive "moving" data continuously and in real time [5, 6]. Obviously, the generation speed and scale of these data have exceeded the processing capacity of traditional distributed systems, because their typical mode is persistent static storage and repeated access of data. The flow computing system should process and analyze the data while it "flows into" the system. In terms of services provided to the outside world, the traditional system provides a single service, the operation plan corresponding to the service is fixed, and the optimization is also implemented before the operation is executed. The services provided by streaming computing system are continuous, and its operation plan is dynamic. Because the scale of real-time streaming data is huge and easy to use, and the scene is complex, it is impossible to enumerate all the services that users may submit. In this case, the real-time computing process is delayed until the demand is submitted. In recent years, there have been many real-time data stream computing systems in the industry. Although there is no integrator similar to, they all have their own characteristics, which is worthy of our study and research. The following is a brief introduction. Several typical streaming computing systems in the industry, including S4, storm and streambase.

S4 (simple scalable streaming system) is a general, distributed, scalable, partition fault-tolerant and pluggable streaming system. Based on the framework, developers can easily develop applications for continuous stream data processing. The design features of S4 are as follows: 1) in order to carry out distributed processing on the cluster composed of ordinary models and do not use shared memory inside the cluster, the architecture adopts the mode. This mode provides encapsulation and address transparent semantics. Therefore, it not only allows large-scale concurrency of applications, but also provides a simple programming interface. The system calculates through the processing unit, the message is transmitted in the form of data events between the processing units, the processing unit consumes events, sends out one or more events that may be processed by other processing units, or directly publishes the results. The state of each processing unit is not visible to other processing units. The only interaction mode between processing units is sending events and consuming events. The framework provides the ability to route events to appropriate processing units and create new processing unit instances. The design pattern conforms to the characteristics of encapsulation and address transparency.

2) In addition to following the pattern, the peer-to-peer cluster architecture also refers to the pattern. In order to simplify deployment and operation and maintenance, so as to achieve better stability and scalability, a peer-to-peer architecture is adopted. All processing nodes in the cluster are equal without central control. This architecture will make the cluster very scalable, and there is no upper limit on the total number of processing nodes in theory. 3) The pluggable architecture system is developed in language and adopts highly hierarchical modular programming. Each general function point is abstracted as a general module as far as possible, and each module can be customized as far as possible. 4) The service-based cluster management layer that supports partial fault tolerance automatically migrates events from failed nodes to other nodes. Unless explicitly saved to persistent storage, when a node fails, the state of processing events on the node is lost.

Storm is a real-time data stream computing system recently opened by twitter. It provides a set of common primitives for distributed real-time computing to process messages in real time and update the database in the way of "stream processing". The main features of storm are as follows: the simple programming model is similar to, that is, it reduces the complexity of parallel batch processing and real-time processing. It supports various programming languages and can use various programming languages on it. To increase the support for multiple languages, you only need to implement a simple communication protocol. Support fault tolerance and manage failures of work processes and nodes. Storm support horizontal expansion: computing is carried out in parallel among multiple threads, processes and servers.

Message processing: Storm ensures that each message can be processed completely at least once. When the task fails, it is responsible for retrying the message from the message source.

Streambase is a commercial streaming computing system developed by IBM, which is used in the financial industry and government departments. The design features of streambase are as follows:

Convenient application interface: Streambase is developed in Java language and provides many operation sets, function sets and other components to help build applications. Users only need to drag and drop the control through the IDE, set the transmission mode and control calculation process, and then they can compile an efficient streaming application. At the same time, Streambase also provides SQL like language to describe the calculation process. By setting the transmission mode and control calculation process, an efficient streaming application can be compiled procedure. At the same time, Streambase also provides SQL like language to describe the calculation process [7, 8].

3 Research Direction of Flow Computing

Stream computing has obvious advantages in real-time distributed processing of massive data. Stream computing mode poses a challenge to the traditional computing architecture. Combined with the research status at home and abroad and the existing data flow computing system, this paper summarizes its research direction, which can be roughly divided into the following points:

1) Overall optimization of performance. Many mature problems that have been solved in the traditional distributed computing mode have not been well solved in the streaming computing system. For example, how to carry out dynamic task allocation and scheduling among multiple nodes of the whole system, and how to make the system make the most full and rational use of system resources when the system resources are limited [9].
2) Complex application requirements. Dynamic demand is one of the key problems that flow computing system is difficult to solve. How to track the time-varying requirements and dynamically adjust the key strategies of the system is a new research direction of flow computing.
3) Efficient cutting and fusion of data streams. Cutting and fusion are the most basic requirements of flow computing. Cutting and fusion will inevitably destroy the time sequence of data flow, and make the flow computing system not suitable for applications sensitive to time sequence, but cutting data flow is the key means to ensure parallel computing. How to ensure the relative order of data items on the basis of parallel processing is also one of the research directions of flow computing system.
4) Research on implementation strategy. At present, in terms of architecture and application, the universality and expansibility of stream processor and stream programming language are not high. The realizability of stream programming language is insufficient, and the stream processor does not realize the autonomy of function [′].

We need more general and efficient strategies to improve the quality of software and hardware and make the stream computing system more practical in service.
5) Improvement of processing capacity. The existing stream computing systems are based on the research of system prototype, and have not been put into the actual production environment on a large scale. Whether the performance of the system can meet the service requirements in the complex massive data stream processing environment needs further research and verification [10].

The data stream processing process can be improved from two aspects: transmission and calculation [11, 12]. With the development of data mining technology, a series of progress has been made in data stream processing algorithm. Literature [13, 14] studies the current situation and future direction of mining. With the increase of data volume, parallel mining will become a mainstream trend; Literature [15, 16] proposed an algorithm for dynamic classification of itemsets by using the characteristics of dynamic changes of data over time, and improved the mining efficiency through classification and pruning strategies. Literature [17, 18] proposed a simplified representation algorithm of fuzzy equivalence classes, which mainly focuses on the errors that reduce the support. The above research has conducted in-depth discussion on the problems existing in the process of data stream mining and achieved certain results, which can improve the efficiency and accuracy of frequent item mining to a certain extent. Although there are many forward-looking research results on data mining algorithms, the real-time transmission and processing technology of data stream needs to be further studied. Therefore, aiming at the improvement of data stream transmission technology to improve the efficiency of data stream processing, a parallel data stream real-time processing mechanism is proposed. Through the flow density processing principle to effectively integrate multi-source data streams, use the channel multiplexing technology to transmit different levels of data streams in time slots, and use the container technology regulation means to change the system function realization mode.

4 Analysis of Data Stream Real-Time Transmission Technology

The business type of power grid is basically dynamic data flow. Due to its real-time, liquidity and variability, its characteristics are difficult to control and track. Therefore, it is necessary to continuously extract historical data from the acquisition side and enter the background calculation and analysis system. Therefore, data stream real-time transmission technology largely determines the service quality and efficiency of power grid communication services. Conventional data stream transmission methods are mainly divided into serial and parallel transmission [19]. Serial transmission is generally used in single machine system, which only provides one channel. The transmission process is relatively slow and easy to cause congestion. Therefore, serial mode is generally not recommended, and the research results are relatively few; distributed parallel is a popular stream transmission mode. It transmits data streams through multiple channels at the same time, which not only alleviates the tense situation of channels, but also prevents the transmission interference between data. At present, there are rtstream and spark mining models in parallel transmission mode. Rtstream model uses a decentralized peer-to-peer architecture to provide the same functions and responsibilities for all nodes. There are a few central nodes responsible for global load balancing, task scheduling and data routing. The streaming data processing capability of the whole framework can also be expanded by adding nodes online. During multi-channel transmission, the failure of any channel will not affect the data transmission and processing results of other channels. It can also support breakpoint continuous transmission and real-time monitoring of data flow operation. Spark streaming is a big data parallel computing framework based on memory computing, which has the characteristics of high speed, high error tolerance, high scalability and so on.

Based on the above research, [20] further optimized the business flow parallel real-time processing technology, combined with the flow density channel reuse method and container technology, proposed an efficient business flow real-time processing mechanism from different dimensions.

Improve the performance of parallel transmission and processing of distribution service data stream. Real time processing mechanism of business flow for power grid source load balance. With the development of new power grid towards diversity and marginality, the accurate transmission and consumption of power between source and load can only rely on the communication network to correctly deliver judgment and decision-making information. In the complex business flow environment, the research on the orderly processing technology of business flow is the key to improve the system efficiency.

The characteristics of data flow determine that the core of data flow algorithm is to design a single pass scanning algorithm, that is, to constantly update a structure representing the data set - summary data structure in a memory space far smaller than the data scale. It makes it possible to quickly obtain approximate query results according to this structure at any time. Give approximate query results in real time. The complexity measurement scales of the algorithm are time complexity and space complexity. The size of the summary data structure should be sublinear at most, and the processing time of each group of data on the stream should not exceed the algorithm complexity. The common methods of generating summary data structure are random sampling, sketching, sliding window, histogram, wavelet and hash. Random sampling method is to extract a small number of samples from the data set that can represent the basic characteristics of the data set. The approximate query results are obtained according to the sample set. A single pass scan data set generates a uniform sampling set. Let the capacity of the sample set be s. at any time n, the elements in the data stream are selected into the sample set with the probability of S/n. If the sample set size exceeds s. Then a sample is randomly removed. The expression efficiency of this method is not high. Later, someone improved the representation of sample set. For elements that appear only once. It is still represented by element code. Save more space. When the sample set overflows, first increase the introduction parameter t to t'. For any one of these elements, the probability T/T'. Then judge whether to subtract 1 with probability $1/T'$. Once the counter value drops to 0. Or if the value of the counter does not decrease after a random judgment, the operation on this element is ended. This method can effectively obtain the list of popular elements in the dataset. Sketch technology assumption $s = (X1 \ldots Xn)$ is a sequence of elements. Where Xi belongs to the value range $d = \{1 \ldots d\}$.

The kth frequency factor (moment of order k) of the sequence s. The frequency factor obtains the statistics of the value distribution of the sequence s. Using the linear hash function requires only memory and space. It has been applied to many database literatures, such as connection size estimation, L1 normal form of estimation variables and complex aggregation of multiple streams. The calculation of F2 is similar to the self-connection of the calculation relationship. The method of estimating the size of two different relational connections to estimating multidirectional connections in response to complex aggregation, and also provides techniques to optimize the segmentation of data domains and estimate each partition independently in order to minimize the overall memory requirements. Histogram technology is to divide a large data set into multiple continuous buckets, that is, small data sets. And each barrel has an eigenvalue. The main histograms are equal width histograms, v-optimized histograms, end bias histograms, etc. The constant width histogram divides the value range data into approximately equal parts. The height of each bucket (i.e. the amount of data contained in the bucket) is averaged. The proposed algorithm uses two important operations (splitting and merging) and two thresholds (upper threshold and lower threshold). V-optimize the histogram to minimize the sum of variances of each bucket. Assume that the value of each element in the dataset is V1, V2 … Vn. Bi represents the average value of the bucket where the element VI is located. Then the value of σ (VI BI) 2 is the smallest. End bias histogram is mainly used for iceberg query. That is, maintain the simple aggregation of data item attributes and query whether the aggregation value exceeds the specified threshold. This iceberg query appears in many applications, such as data mining, data warehouse, information acquisition, sales analysis, replication detection and clustering. Motwani introduced the random decision algorithm for frequency counting and iceberg query of data streams. The algorithm is adaptive sampling and maintains a sampling of different frequency terms. When a data item exists in the sampling, its frequency is increased, and the low-frequency data items are deleted periodically. The problem of constructing summary data structure under sliding window model is. When new data continues to arrive. When old data expires. How to process expired data. Make the query results reliable. The basic window method is divided into k equal width sub windows in chronological order. Each basic window contains w/k elements. And a small structure represents the characteristics of the basic window. If the window contains elements that have expired. Delete the small structure representing this basic window. Users can get approximate query results based on these small structures that have not expired. The following algorithm describes the sample set uniformly sampled on the sliding window. At any point in time, the elements in the N stream are added to the sample set in probability. When an element is selected into a sample set. An alternative element needs to be determined. To replace this element when it expires. Because future data cannot be predicted in the data flow. Therefore, only select a number randomly from [N + 1… N + w] as the timestamp t of the candidate element. When time point t is reached. This alternative element is finally determined. Any element in the sample set. "Chain" with one alternative element. Once the element expires. Immediately replace it with the next element on the chain.

5 Real Time Processing Technology of Business Flow

In the electric communication network, the concentrator or gateway equipment integrates the bandwidth, rate and service parameters of the data stream output by the lower terminal equipment. All heterogeneous data are unified in data format in network equipment, forming data blocks of different sizes, which are cached in the middle layer. The collection terminals of the power grid service system are scattered and have many points, and the collection traffic density and bandwidth of different services are different [21, 22]. In order to make full use of the stream channel resources, the utilization rate is maximized according to the stream density channel multiplexing technology [23]. The data services passed by the power grid concentrator or gateway equipment include various services collected in parallel to form a serial data link, and the gateway data will be uploaded to the upper background distributed database [18]. By reorganizing the chain data into virtual channels according to the size of data blocks, the basic data blocks transmitted in each channel shall be consistent as far as possible. In the real-time processing of data stream, the main function of channel multiplexing is to make full use of transmission channel resources, carry data in multiple channels, improve the data transmission rate, and have the advantages of independence and strong anti-interference ability between data blocks [20]. The classification process is carried out simultaneously in parallel. After classification according to the density, the data streams are summarized in different stream sets according to the density [24, 25]. In the s set, the data flow is divided into equal size data blocks, and the data blocks with high density and low density are combined to form a standard basic data block to avoid channel congestion. From the stream channel multiplexer to the distributed database, the data storage link is mainly completed, and the time period is divided into small time slots $= 1\ \mu s$. After passing through the channel multiplexer, multiple parallel storage channels are transferred out, and the low-density S1–S3 is inserted into the large time slot S4–S5 as much as possible to form a data block of equal size, which can complete the I/O action at the maximum rate, and the storage channel is saturated on the time slot. Data processing enters the stream memory in FIFO order. The mapping relationship between storage channel and distributed database is many to one, so there is a problem that multiple channels are written in parallel at the same time. Virtual channel will be used for docking at the database end, which will not affect the data writing rate. The storage speed is improved. There is no need to set too many cache devices and increase the system power consumption [22]. There is a big problem in using channel multiplexing technology, that is, the repeated matching between data streams will lead to bit error in the transmission process. Considering the safe storage of data streams, we should ensure that the bit error rate of data streams reaches more than 90%. Distributed memory mainly serves various business departments in the background [26, 27]. It applies the data functions that businesses care more about, virtualizes the data functions, and forms a storage container. The business application system can directly realize the data application by calling the function container. Due to the huge amount of data in the database, in order to improve the calling efficiency and data flow quality, container based stream preprocessing technology is adopted. Docker is similar to virtualization technology. They prefer different application scenarios. Due to the variability of stream data, the process of stream data processing belongs to a dynamic process, and stream data is not easy to store statically.

Virtualization technology is more suitable for virtualizing static resources and forming resource pools, so as to centralize servers and other devices. Through the combination of stream memory and virtual control platform, the platform includes multiple classification virtual containers and API interfaces. Each distributed database corresponds to a virtual function container, which is convenient for direct mapping to data flow. Through the call of the container, the data I/O of wrong instructions is avoided, and the data is often read and written, which is easy to increase the risk of data security or data loss. The called outflow is transmitted to the corresponding application service system after denoising, feature extraction, separation and other steps in the preprocessing subsystem.

6 Conclusion

This paper summarizes the current situation of distributed data stream processing technology at home and abroad, mainly including the definition of data stream, data stream model, real-time transmission technology of data stream, data stream processing technology and so on. This paper provides an analysis of the current situation at home and abroad for the distributed data flow processing of power grid regulation, which is conducive to promote and improve the promotion and application of the distributed data flow processing technology and framework of power grid regulation. In the future, we will further apply these technologies to the application scenario of real-time processing of power grid distributed business data flow. At the same time, aiming at the shortcomings of existing data streams, a distributed stream data resource processing model and framework suitable for the field of power grid dispatching are proposed.

Acknowledgements. This paper is supported by the State Grid Anhui Electric Power Co., Ltd. Science and technology project "Research on real-time transparent access technology to improve the perception ability of power grid operation state".

References

1. Babcock, B., Babu, S., Datar, M.: Models and issues in data stream systems. In: Proceedings of the 21st ACM SIGACT SIGMOD-SIGART, pp. 1–16 (2002)
2. Madden, S., Franklin, M.J.: Fording the stream: an architecture for queries over streaming sensor data. In: Proceedings of the 18th International Conference on Data Engineering, pp. 555–566. Morgan Kaufmann Publishers, San Jose (2002)
3. Zeng, Y.F., Yang, X.J.: The optimized organization method of stream on image stream processor. J. Comput. Sci. **31**(7), 1092–1100 (2008)
4. Kuo, K., Rabbah, R.M., Amarasinghe, S.A.: Productive Programming Environment for Stream Computing. Computer Science and Artificial Intelligence Laboratory, Massachusetts Institute of Technology
5. Babu, S., Widom, J.: Continuous queries over data streams. SIGMOD Rec. **30**(3), 109–120 (2001)
6. Arasu, A., Babcock, B.: Stream: The Stanford Stream Data Manager (Demonstrate ion Description), San Diego, CA, United States, p. 665 (2003)
7. Chandrasekaran, S., Franklin, M.J.: Streaming queries over streaming data. In: Proceedings of the International Conference on VLDB, pp. 203–214 (2002)

8. Terry, D., Goldberg, D., Nichols, D.: Continuous queries over append-only databases. In: SIGMOD (1992), pp. 321–330, June 1992
9. Shixuan, S.A., Wang, S.: Introduction to Database System, pp. 220–267. Higher Education Press (2000)
10. Tatbul, N., Cetintemel, U., Zdonik, S.B.: Load shedding in a data stream manager. In: Proceedings of the 29th International Conference on Very Large Data Bases, Berlin Germany, pp. 309–320, September 2003
11. Xiong, J.: Research and Application of Key Technologies of Distributed Stream Processing. University of Electronic Science and Technology, Chengdu (2017)
12. Qi, J., Qu, Z., Lou, J., et al.: A kind of attribute entity recognition algorithm based on Hadoop for power big data. Power Syst. Protect. Control **44**(24), 52–57 (2016)
13. Wang, H., Fu, Y.: Heterogeneous network convergence-research development status and existing problems. Data Commun. **2**, 18–21 (2012)
14. Gu, X.: Research and Implementation of Key Technologies of Distributed Streaming Computing Framework. Beijing University of Posts and Telecommunications, Beijing (2012)
15. Jiang, D., Zheng, H.: Research status and developing prospect of DC distribution network. Autom. Electr. Power Syst. **36**(8), 98–104 (2012)
16. Chen, Z.: Data Stream Clustering Analysis and Anomaly Detection Algorithm. Fudan University, Shanghai (2009)
17. Liu, K., Sheng, W., Zhang, D., et al.: Big data application requirements and scenario analysis in smart distribution network. Proc. CSEE **35**(2), 287–293 (2015)
18. Yang, X., Zeng, L., Yu, D.: The optimized approaches of organizing streams in imagine processor. Chin. J. Comput. **31**(7), 1092–1100 (2008)
19. Yang, Y., Han, Z., Yang, L.: Survey on key technology and application development for data streams. Comput. Appl. Res. **11**, 60–63 (2005)
20. Lu, J.: Introduction to Distributed Systems and Cloud Computing. Tsinghua University Press Xinhua News Agency, Beijing (2013)
21. Zhu, F., He, Y.: Theory and Design of Scheduling Algorithm in Parallel Distributed Computing. Wuhan University Press, Wu Han (2003)
22. Yu, Y., Lv, Z., Qi, G.: Research distributed large-scale time series data management platform. Power Syst. Protect. Control **44**(17), 165–169 (2016)
23. Chang, G., Hao, J., Liu, B., et al.: Research and development of intelligent and classified collection system for electric power dispatching and control information. Power Syst. Protect. Control **43**(6), 115–120 (2015)
24. Arasu, A., Babu, S., Widom, J.: The CQL continuous query language: semantic foundations and query execution. VLDB J. **15**(2), 121–142 (2006)
25. Cortes, C., Fisher, K., Pregibon, D., et al.: Hancock: a language for analyzing transactional data streams. ACM Trans. Program. Lang. Syst. (TOPLAS) **26**(2), 301–338 (2004)
26. Zhang, T., Liang, S., Gu, J.: Overview of the distribution and utilization big data application. Electr. Meas. Instrum. **54**(2), 92–99 (2017)
27. Zhang, X., Yang, G., Zhao, G.: Research on automatic diagnosis and analysis technology of distribution network status based on global large power data. Electr. Meas. Instrum. **56**(16), 111–115 (2019)

Modern Information Technology and Application

NVLSM: Virtual Split Compaction on Non-volatile Memory in LSM-Tree KV Stores

Zhutao Zhuang[1(✉)], Yongfeng Wang[1], Shuo Bai[2], Yang Liu[2], Zhiguang Chen[1], and Nong Xiao[1]

[1] Sun Yat-sen University, Guangzhou, China
zhuangzht@mail2.sysu.edu.cn, yongfeng.wang@nscc-gz.cn
[2] School of Computer Science, National University of Defense Technology, Changsha, China
liuyang_@nudt.edu.cn

Abstract. Log-Structured Merge tree is a write optimized persistent storage engine consisting of memory buffer and multiple layers of disk files. Log-Structured Merge tree is built for block devices and suffers from write stalls due to its frequent internal L0–L1 compaction operation. Emerging storage hardware non-volatile memory (NVM) brings opportunities for optimization of LSM-tree with its byte-addressability, high bandwidth, low latency. In this paper, we introduce NVLSM, a novel LSM-tree design based on hybrid storage of NVM with SSD and NVM with HDD to improve write throughput of Log-Structured Merge tree. Experimental results show that NVLSM achieves 1.11x higher random write throughput compared to the baseline model, and the read performance is comparable.

Keywords: LSM-tree · Non-volatile memory · Key value storage

1 Introduction

1.1 A Subsection Sample

Persistent key-value store plays an important role in modern business application, cloud storage and IoT storage. B-tree and Log-Structured Merge tree (LSM-tree) are two typical kind of key-value storage engine which features read-friendly and write-friendly respectively. Well-known LSM-tree based key-value storage engine such as LevelDB, RocksDB [5], Cassandra [10] are optimized to support write-intensive workload.

LSM-tree achieve high write throughput by batching writes into memory buffer and flush memory buffer into persistent storage when memory buffer exceeds size limit. Data on persistent storage for LSM-tree is organized as multiple sorted levels. To remove duplicate keys and maintain sorted order of keys, LSM-tree will do compaction work at background. LSM-tree suffers from write amplification and write stall due to its compaction work.

Emerging non-volatile Memory (NVM) such as phase change memory and 3D XPoint brings new opportunities to performance improvement of storage systems. Data

© The Author(s), under exclusive license to Springer Nature Switzerland AG 2023
L. C. Tang and H. Wang (Eds.): BDET 2022, LNDECT 150, pp. 97–106, 2023.
https://doi.org/10.1007/978-3-031-17548-0_9

can be persisted on NVM and it can be accessed like DRAM with byte-addressability of NVM. Bandwidth and latency of NVM is comparable with DRAM while NVM has large large capacity and higher density than DRAM.

In this work, we propose Non-volatile Virtual Split Compaction Log-Structure Merge tree (NVLSM), an novel LSM tree design that exploits byte-addressability and high bandwidth of non-volatile memory to increase random write throughput and mitigate write stall of LSM tree.

NVLSM reduces write amplification and improve write throughput by storing data files of Level0 and Level1 on NVM and doing virtual compaction instead of physical compaction that costs bandwidth of NVM and increases latency of user operation.

The main contributions of this works are as follows:

- We introduce NVLSM, an novel LSM-tree design that is built on NVM-SSD and NVM-HDD hybrid storage layer that utilize high bandwidth and low latency of NVM.
- We introduce new storage format for data file stored on NVM and virtual compaction scheme on Level0–Level1 compaction with small reading overhead and without actual data re-writing, which improves write throughput dramatically of LSM-tree.
- We implement NVLSM based on LevelDB [13], a well-known open source LSM-based key-value store. Our experimental results shows that NVLSM shows 1.11x write throughput improvement on SSD and 1.04x write throughput improvement on HDD compared to baseline model.

2 Background

2.1 Log-Structured Merge Tree

LSM-tree is a write optimized persistent key-value store proposed by O'Neil et al. [12]. LSM-tree achieves high write throughput by batching writes in memory buffer first, then flushing the memory buffer into the disk when the size of memory buffer exceeds the size limit.

Here we explain some key concepts of a popular LSM-tree implementation LevelDB and illustrate how key-value data is organized and maintained internally in memory and persistent storage media. Figure 1 shows the overall storage structure of LevelDB. In LevelDB, the memory buffer which used for batching writes is called MemTable and is implemented as skiplist. The MemTable will be converted from mutable to immutable once the MemTable is full and a new MemTable will be created to serve upcoming write requests. The immutable MemTable will then be compacted into a file called SSTable in persistent storage media, in which keys are stored in sorted order.

SSTables are stored in multiple levels in persistent storage media like SSD and HDD. SSTables generated from newly flushed immutable memtable will be put at level0, at which they may have overlapping key ranges with each other. Then the key-value pairs in SSTables will be gradually moved from level0 to deeper levels (L1, L2,... Ln) during compaction work done by background threads. SSTables in deeper levels except level0 are strictly sorted, meaning key ranges of SSTables do not overlap.

The procedure of compaction is as follow:

(1) One SSTable will be picked from Li, and multiple SSTables in Li+1 whose key ranges overlap with that from Li will also be picked as compaction data input. (2) Key-value pair data from input SSTables will be fetched into memory and merged, sorted and written into new SSTables in Li+1 except those marked as deleted and outdated. (3) Those SSTables involved in compaction will be deleted from the LevelDB.

Since key ranges of SSTables in L0 are allowed to overlap, all SSTables in L0 whose key ranges overlap with the first picked SSTable in L0 will also be involved in compaction as key-value input source.

The main purpose of compaction is to maintain sorted order of keys in each level so that lookup operation is accelerated by binary search and keep the size of each level of SSTables from growing up indefinitely. However, this compaction process will cause the same key-value data pair re-written multiple times, resulting in write amplification and increased write latency, and lower system throughput.

The procedure of lookup operation is as follows: (1) The lookup key will be searched in mutable MemTable. (2) If not found in mutable memtable and there is an immutable MemTable not yet been flushed into SSTable, then lookup key will be searched in immutable MemTable. (3) If there is no key match in MemTable or immutable MemTable, LevelDB will search from L0 to Ln until a key match or return not found.

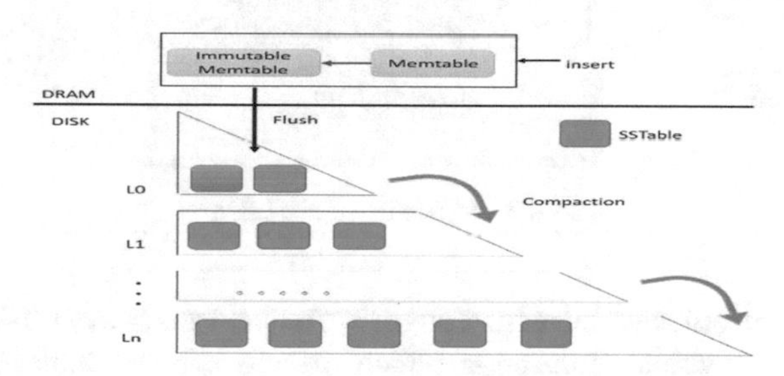

Fig. 1. Structure of LevelDB

2.2 Non-volatile Memory

Non-volatile memory such as phase change memory [14], 3D XPoint [7] and spin transfer torque MRAM [6] brings new opportunity to improve storage system performance. Non-volatile memory is byte-addressable, non-volatile, and fast. NVM has a higher capacity and larger density than DRAM, and it is cheaper than DRAM, which makes it suitable to store hot data. Read and write latency of NVM is 100x lower than SSDs. NVM provides 10X higher bandwidth than SSD [8].

3 Non-volatile Virtual Split Compaction Log-Structured Merge Tree (NVLSM)

3.1 Overall Architecture

NVLSM achieves high write throughput and low write latency by efficient use of high bandwidth and low latency, byte-addressability of NVM.

Figure 2 shows the overall architecture of NVLSM. SSTables at level 0 and level 1 are stored on NVM to utilize the high bandwidth of NVM. SSTable format is also redesigned so fast random access ability and byte-addressability of NVM can be fully utilized. As mentioned before, compaction of LSM-tree takes place mostly on the upper level of LSM-tree, especially on L0–L1 compaction, which costs bandwidth and increases the latency of user operation, decreasing system throughput. Almost all SSTables on L0 and L1 will be involved in compaction since L0 allows overlapping key ranges of SSTables, which costs massive CPU cycles and bandwidth of storage media.

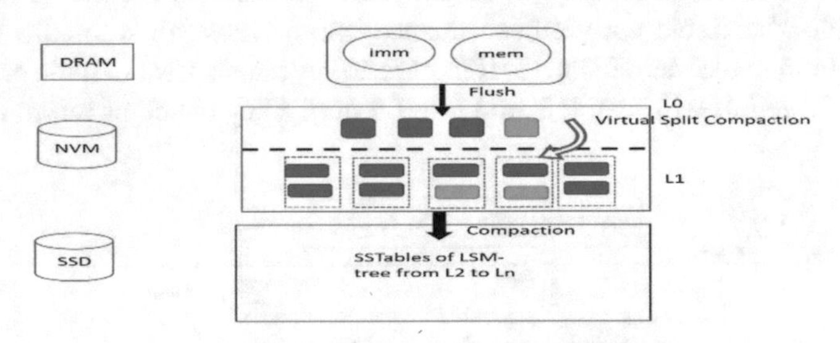

Fig. 2. Structure of NVLSM

To mitigate performance decrease brought by frequent and heavy L0–L1 compaction, NVLSM introduces virtual compaction, which virtually splits SSTables on L0 into multiple segments and append these segments to virtual SSTable on L1 instead of reading a massive amount of data, merging them, and then writing new data back to new SSTables like standard compaction. Virtual compaction of NVLSM does not write data to NVM and only incurs a small reading overhead to split the SSTables on L0 into multiple segments so virtual compaction is suitable to work well under intensive write workload.

3.2 Table Format

The storage format of SSTable on LevelDB is optimized for block devices such as HDD and SSD, which consists of multiple kinds of blocks, named data block, index block, and meta index block. Flushing blocks to disk while building SSTables will go through a deep file system software stack, bringing significant extra overhead. With DAX support of the operating system and PMDK library, we can access NVM bypassing the traditional file system call stack. To fully utilize byte-addressability of NVM, we propose a new storage format for SSTables on NVM, which we call NVSSTable. Figure 3 illustrate the format of the NVSSTable and LevlDB SSTable.

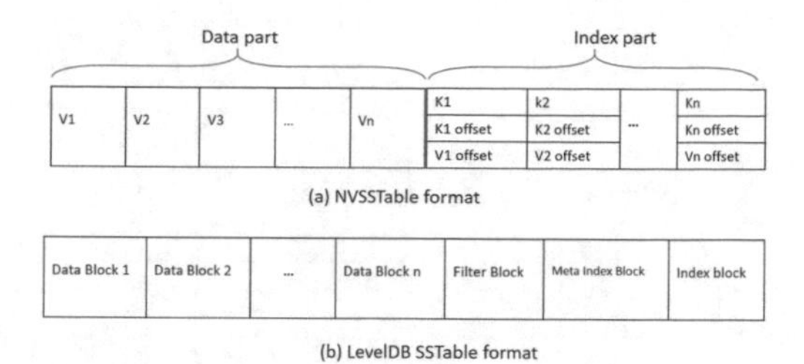

Fig. 3. NVSSTable and LevelDB SSTable

NVSSTable consists of two parts, namely the index part and the data part. The index part stores array of key units. The Key unit consists of key length, key, key offset, and data offset. NVTable separates key and value to utilize byte-addressability of NVM, lookup operation will do binary search in the index part first and then return the corresponding value of the key given the data offset of the key when the key is located in the index part. Separation of key and value will shorten search range and brings less access times of NVM in one lookup operation.

Besides locating the position of the value in the NVSSTable, data offset is also used in the L0–L1 virtual split compaction of NVLSM on NVM along with key offset to split the NVSSTable virtually. The virtual split compaction will be discussed in 3.3.

3.3 Virtual Split Compaction

Virtual split compaction is a compaction technique that virtually splits the NVSSTable at L0 into multiple segments and appends the newly generated segments to the end of the virtual NVTable at L1. NVTable is a logical collection of segments whose key range overlaps. Virtual split compaction splits the NVSSTables with small read overhead instead of original heavy read-merge-write compaction.

The steps of virtual split compaction at L0–L1 is as follow: (1) NVLSM picks one NVSSTable from L0 as a candidate like LevelDB, and all the NVSSTables at L0 whose key ranges overlaps with that of the candidate will also be picked. (2) The NVTables at L1 whose key ranges overlap with candidates at L0 will be selected to receive segments split from NVSSTables at L0. A segment is a logical view of the continuous part of NVSSTable, which includes the index part and the corresponding data part. (3) Before splitting NVSSTables into segments, we will virtually split NVTable into two NVTable whose file size is biggest among all the candidates at L1. (4) Then, we iterate all the NVSSTables of candidates at L0 and start splitting each one into multiple segments whose key ranges are contained by the NVTables at L1. The segments will be appended to the end of the corresponding receiving NVTable. This way, we could guarantee that the key range of NVTables at L1 is strictly sorted. (5) NVTables at L1 will gradually accumulate segments until its size exceeds NVTable size limit and triggers compaction from L1 to L2, which will compact all the segments into SSTables to L2.

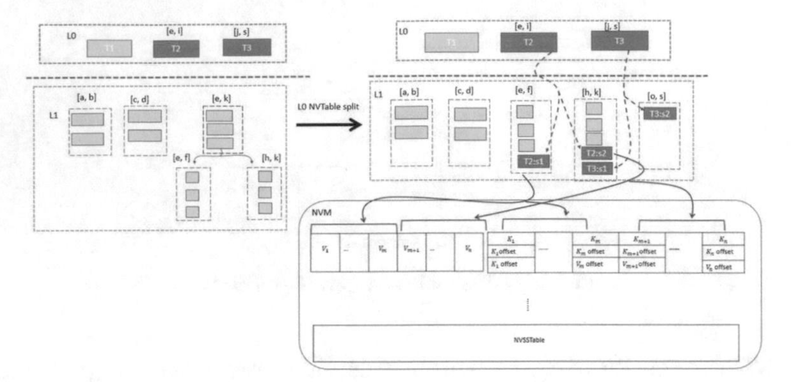

Fig. 4. Virtual split compaction: an example

For example, in Fig. 4, the NVSSTable T2 and T3 at L0 are picked as candidates to be virtually compacted to L1. NVTable V2 at L1, whose key range overlaps with candidates at L0, is selected to receive the splitting segments from T2 and T3. Since the total key-value size of V2 is the biggest among all the NVTables candidates, V2 will be split into two NVTables, V4 and V5, and cut key ranges of V2 into half. Then the NVSSTable T2 splits into segments T2:S1 and T2:S2 whose key ranges are within [d, f] and [h, k] and T2:S1 and T2:S2 will be appended to the back of V4 and V5. The split process of NVSSTable T3 is similar to that of T2 except that T3:S2 will be appended to an newly created empty NVTable V6 since key range [o, w] of T3:S2 does not overlap with key range of V5. NVLSM does not enlarge the key range of NVTable by receiving out-of-range segments because we want to keep the range of the NVTable as small as possible since NVSSTables will grow size by gradually accumulating upcoming segments from NVSSTable at L0.

Virtual split compaction will accumulate keys within the same range in the same NVTable for a relatively long time to curb the performance degrade that frequent L0–L1 compaction cause. Hence the write throughput will increase. Also, segments whose key range overlaps are likely to contain keys that are overwritten multiple times, which will be deleted during actual L1–L2 compaction, reducing space amplification.

4 Evaluation

4.1 Experiment Setup

All experiments are run on a machine with one Intel(R) 2.30 GHz 40 core processors and 128 GB memory. The kernel version is 64-bit linux 3.10 and the operating system is CentOS7. The experiments use three storage devices which is 1.4T NVMeSSD, 512G HDD and 2x 128G Optane NVMs.

We compare NVLSM with LevelDB and LevelDB with SSTables at L0 and L1 stored on NVM which is called LevelDB_NVM. The size threshold of single NVTable at L1 is set to 20 MB and total NVTable size threshold is set to 100 MB instead of 10 MB to accumulate enough data of segments to trigger compaction from L1 to L2.

4.2 Benchmark Results

In this section, we evaluate the overall performance of LevelDB, LevelDB_NVM and NVLSM using db_bench of LevelDB. The read and write performance are evaluated by inserting 1 GB KV items with different value size in sequential order and in random order separately.

Write Performance Comparison
Figure 5 shows the random write throughput and sequential write throughput of three KV stores as a function of five different value size of inserting entries. Throughput is measured in MB/s. NVLSM improve random write throughput and sequential throughput significantly over LevelDB and LevelDB_NVM on large value size from 1024 byte. The write performance improvement of NVLSM over LevelDB_NVM ranges from 0.16x to 1.11x. The write throughput improvement of NVLSM comes from virtual split compaction which avoids the frequent L0–L1 compaction on LevelDB and the NVSSTable format designed for fast segment generation.

For small value size entries, NVLSM achieve better or comparable results than LevelDB and LevelDB_NVM.

Notice that random write throughput and sequential write throughput of LevelDB_NVM is lower than LevelDB for L2–Ln storage media NVMeSSD since NVMeSSD is relatively fast and data migration from NVM to SSD on L1–L2 compaction will happen while there is no physical data reading and writing for trivial move compaction on model LevelDB.

Read Performance Comparison
Figure 6 and Fig. 7 shows sequential read performance comparison and random read performance comparison with L2–Ln storage media SSD and HDD. The result show that read latency of NVLSM is 7.5% higher for sequential pattern read on with base storage media SSD and HDD, which is a acceptable result with significant write throughput improvement. NVLSM uses heap to speed up fetching the smallest elements across all segments in one NVTable which also contributes to the comparable sequential read performance.

Read latency for random read pattern shows that NVLSM achieve comparable or better results for small value size of keys performs poor when the value size is 4 KB because number of keys with larger value size in one NVTable is less compared to keys with smaller size, thus percentage of keys on NVM is larger than keys with smaller value size and doing binary search on all segments of one NVTable is more expensive than that on SSTables which contributes more to the latency of read operation.

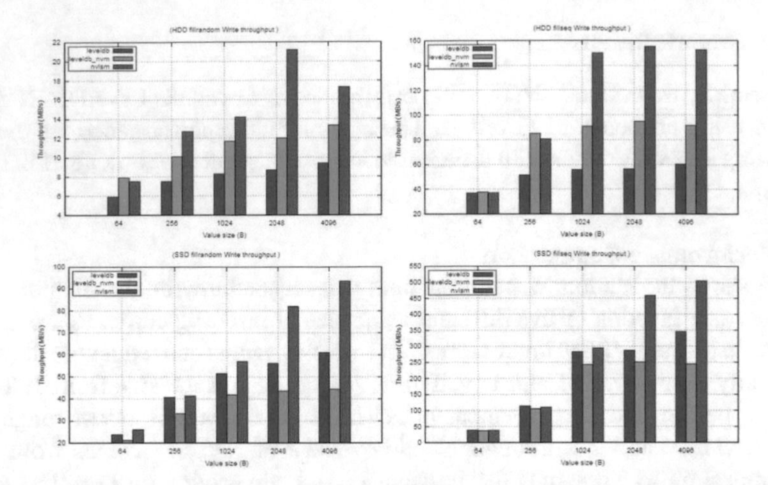

Fig. 5. Write throughput comparison with different storage media and write pattern

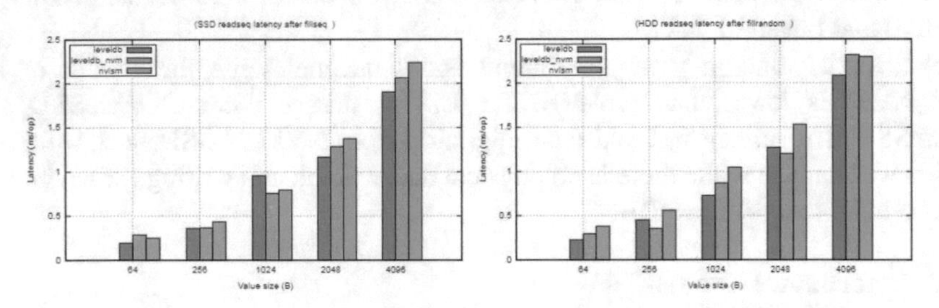

Fig. 6. Sequential read performance comparison with HDD and SSD

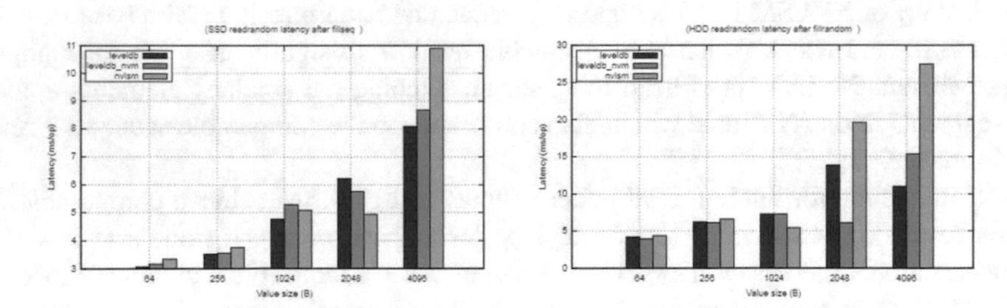

Fig. 7. Random read performance with HDD and SSD

5 Related Work

5.1 Compaction Optimization

LSM-tree suffers from write amplification, read amplification and space amplification due to its internal compaction and multi-level structure. Optimization techniques to improve the performance of LSM-tree by optimize compaction procedure have been extensively studied. LDC [3] introduces lower level data driven compaction method for SSD-oriented LSM-tree store to reduce tail latency and improve write throughput.

5.2 LSM-Tree on NVM

Occurrence of NVM opens new opportunity to improve performance of current KV store. There has been growing interests in utilizing NVM to bring high performance and overcome shortcoming of LSM-tree based store. MatrixKV [15] uses NVM to store all SSTables on Level0 and does fine grained L0–L1 column compaction to reduce write stall and write amplification. NoveLSM [9] exploits NVM to store extra MemTable and Immutable MemTable on NVM to deliver high throughput. NoveLSM allows direct update to MemTable stored on NVM to reduce number of compaction. MyNVM [4] uses NVM as a block device to reduce DRAM footprint in SSD based KV store.NVMRocks [11] builds NVM-aware RocksDB by maintaining persistent MemTable on NVM.

6 Conclusion

In this paper, we present NVLSM, a novel LSM-tree based KV store that is designed for hybrid storage media system which consists of byte-addressable devices NVM and block devices HDD or SSD. NVLSM introduce NVSSTable and virtual split compaction to utilize byte-addressability to avoid frequent L0–L1 compaction and thus help improve system write throughput. The experimental results show that NVLSM achieve higher random write throughput and sequential write throughput compared to baseline model.

Acknowledgments. This work is supported by The National Key Research and Development Program of China (2019YFB1804502). NSFC: 61832020. Guangdong Natural Science Foundation (2018B030312002) the Major Program of Guangdong Basic and Applied Research: 2019B030302002. Supported by the Program for Guangdong Introducing Innovative and Entrepreneurial Teams under Grant NO. 2016ZT06D211.

References

1. 2016: Persistent Memory Development Kit, 27 May 2021. https://pmem.io/pmdk
2. Caulfield, A.M., De, A., Coburn, J., Mollow, T.I., Gupta, R.K., Swanson, S.: Moneta: a high-performance storage array architecture for next-generation non-volatile memories. In: 2010 43rd Annual IEEE/ACM International Symposium on Microarchitecture, pp. 385–395 (2010). https://doi.org/10.1109/MICRO.2010.33
3. Chai, Y., Chai, Y., Wang, X., Wei, H., Bao, N., Liang, Y.: LDC: a lower-level driven compaction method to optimize SSD-oriented key-value stores. In: 2019 IEEE 35th International Conference on Data Engineering (ICDE), pp. 722–733 (2019). https://doi.org/10.1109/ICDE.2019.00070
4. Eisenman, A., et al.: Reducing DRAM footprint with NVM in Facebook. In: Proceedings of the Thirteenth EuroSys Conference, pp. 1–13 (2018)
5. Facebook, RocksDB: RocksDB: a library that provides an embeddable, persistent key-value store for fast storage (2019). https://github.com/facebook/rocksdb

6. Huai, Y.: Spin-Transfer Torque MRAM (STTMRAM): Challenges and Prospects (2008)
7. Intel and micron produce breakthrough memory technology (2019). https://newsroom.intel.com/news-releases/intel-and-micron-produce-breakthrough-memory-technology/. Accessed 27 May 2021
8. Condit, J., et al.: Better I/O through byte-addressable, PersistentMemory (n.d.)
9. Kannan, S., Bhat, N., Gavrilovska, A., Arpaci-Dusseau, A., Arpaci-Dusseau, R.: Redesigning LSMs for nonvolatile memory with NoveLSM. In: 2018 {USENIX} Annual Technical Conference ({USENIX}{ATC} 2018), pp. 993–1005 (2018)
10. Lakshman, A., Malik, P.: Cassandra: a decentralized structured storage system. ACMSIGOPS Oper. Syst. Rev. **44**(2), 35–40 (2010)
11. Li, J., Pavlo, A., Dong, S.: NVMRocks: RocksDB on nonvolatile memory systems (2017)
12. O'Neil, P., Cheng, E., Gawlick, D., O'Neil, E.: The log structured merge-tree (LSM-tree). Acta Informatica **33**(4), 351–385 (1996)
13. Ghemawat, S., Dean, J.: LevelDB: a fast key value storage library (2016). https://github.com/google/leveldb. Accessed 27 May 2021
14. Wong, H.S.P., et al.: PhaseChange memory. Proc. IEEE **98**(12), 2201–2227 (2010). https://doi.org/10.1109/JPROC.2010.2070050
15. Yao, T., et al.: MatrixKV: reducing write stalls and write amplification in LSM tree based KVStores with matrix container in NVM. In: 2020 USENIX Annual Technical Conference (USENIX ATC 2020), pp. 17–31. USENIX Association (2020)

Evaluation of the Performance Character of SPIRIT Value Through Pancasila Education During the Covid-19 Pandemic

Erma Lusia[1](✉), Arcadius Benawa[2], Alfensius Alwino[3], Iwan Irawan[3], and Dian Anggraini Kusumajati[4]

[1] Tourism Department, Faculty of Economics and Communication, Bina Nusantara University, Jakarta, Indonesia
ermalusia@binus.ac.id

[2] Character Building Development Center, Mass Communication Department, Faculty of Economics and Communication, Bina Nusantara University, Jakarta, Indonesia
aribenawa@binus.ac.id

[3] Character Building Development Center, Computer Science Department, School of Computer Science, Bina Nusantara University, Jakarta, Indonesia
{alfensius.alwino, iwan.irawan}@binus.ac.id

[4] Character Building Development Center, Information System Department, School of Information System, Bina Nusantara University, Jakarta, Indonesia
diananggi@binus.ac.id

Abstract. In the midst of efforts to deal with the ongoing COVID-19 attack, values education has challenges as the impact of COVID-19. The change in the teaching and learning process from onsite to online is a challenge for teachers of value education courses. In this situation the purpose of this study is to observe the extent of students' understanding of Pancasila Education towards Value Education, especially, SPIRIT Values of the Students. The method used a quantitative research method which is carried out by distributing questionnaires to 150 respondents as simple random sampling and the data was processed by the SPSS 22 program, then completed by using path analysis. The results showed that Pancasila Education has a significant influence, which is indicated by the significance value in the anova table and the coefficient table of 0.000 which means it is smaller than the 0.05 probability. The result of the study found that Pancasila education had a significant influence on the Value Education of the students. It is recommend that Pancasila Education should be delivered more intensively and seriously so that instilling values, especially SPIRIT values to the students as the young generation through Pancasila Education is more effective. So that the students are not only having an understanding of Pancasila but can also implement the SPIRIT values as proclaimed by Binus University as the core values. Although learning has undergone a change in pattern from onsite to onsite due to the ongoing pandemic, with various obstacles faced, students still have to maintain values such as SPIRIT, namely Perseverance, Innovation, Respect, Integrity, and Team-Work.

Keywords: Pancasila education · Value education · SPIRIT character · COVID-19 pandemic

© The Author(s), under exclusive license to Springer Nature Switzerland AG 2023
L. C. Tang and H. Wang (Eds.): BDET 2022, LNDECT 150, pp. 107–116, 2023.
https://doi.org/10.1007/978-3-031-17548-0_10

1 Introduction

The current unpredictable change has taught various parties to anticipate. The latest is the Covid-19 attack which has had a major impact on the field of education. Increasing immunity, such as giving vaccines, continues to be carried out in an effort to protect health from various viruses, not only Covid-19 but also from various diseases caused by climate change because even though the impact is slower but more dangerous in worsening human health [1].

The COVID-19 pandemic has affected the student teaching and learning process, including the number of closed classes and the shifting of the learning process to online [2]. Even so, the lockdown still provides opportunities for lectures and students to continue to interact, teachers give assignments to students via the internet, lectures deliver it through live conferences using various applications such as Zoom, Google meet, Facebook, Youtube, and Skype etc. [3]. This is certainly a challenge for teachers in adjusting the way of learning for students.

Character education in Indonesia aims to build the potential of students to become human beings who have the skills, desires, and good morals as stated in the Law on the National Education System [4]. Education prepares students to become individuals who have knowledge, skills, attitudes, values, and behaviors that will have a positive impact on students' daily lives in society [5]. Pancasila education is one way to instil a moral and broad-minded personality in the life of the nation and state. Therefore, Pancasila education needs to be provided at every level of education, from elementary, secondary to tertiary levels [6]. The use of Pancasila as the element of research can not be separated from the purpose to elevate the humanistic element [7]. Currently the curriculum must also prioritize value education for students so that they can maintain the values in themselves, they can determine what is right and wrong, good or bad and justice or injustice [8]. Value education is a process that starts at home and continues in society and in formal educational institutions [9]. Values education can be delivered to students in various ways, but the main purpose of value education to students in formal educational institutions is to make them understand and be able to implement the importance of values; good daily behavior and attitude; and be able also to contribute to society through social responsibility and good ethics [10], no exception in relation to value education at Bina Nusantara University, namely SPIRIT, such as Striving for the excellence, Perseverance, Innovation, Respect, Integrity, and Team-Work, which are the core values at Bina Nusantara University.

The last few years have shown the lack of student interest in Pancasila [6]. Other researcher observed that the attitudes seen from some people and students showed attitudes that not applied Pancasila values in everyday life. Therefore, it is very necessary for Pancasila education to be taught at the school and college level [11]. Various factors that cause the decline in knowledge and understanding of Pancasila among students must be explored and the best solution sought to re-strengthen knowledge and understanding of Pancasila ideology among students. The impact of the COVID-19 pandemic which has limited student social interactions, as well as drastic changes in the way of learning have become a challenge for teachers to continue to teach values to students. Based on that phenomena, the research aims to reveal the significance of Pancasila education on the value education of the students referring to the Binus University core values, which well known as SPIRIT.

2 Literature Review

Pandemic covid-19 has a bad impact on the learning system of students, this condition makes many teachers and students hope for innovation to ensure online education is maintained during the pandemic, the education sector is struggling to survive the crisis with a different approach, by using digital methods to overcome the challenges created by the pandemic [11]. Distance learning is one solution to continue the teaching and learning process that has been disrupted due to COVID-19, however, the lack of network infrastructure, computers, and internet access are the challenge for distance learning [12]. Previous research has found that distance learning can be an alternative to traditional education, the combination approach between online and traditional education can support the education process to be more effective [13].

Other researcher conclude that Pancasila is a reflection of the educational character in building the national identity of Indonesia as a great nation, modern, dignified and civilized [12]. Pancasila is also the guidelines for people to act and have to be understood and implemented in daily life [13]. At the college level, students need to grow in their environment to learn how to feel empathy for others, to share, to learn rationality, spirituality, competence in technology; as well as communication skills, and other fields that will support their life in society [10]. Previous study showed that students can implement the value of Pancasila. It meant that they understood the values of Pancasila and be able to implement in their daily lives [17]. Students' integrity is also increasingly formed by participating in Pancasila Education learning through habituation and role models. Habituation of learning activities and role models come also from the teacher concerned [18]. Value education plays a role in assessing the criteria for values consisting of awareness, emotion, excitement and integration of individual's behavior, and The main purpose of value education is to instill good behavior so that it becomes permanent behavior in students' daily lives [16]. Value education prepares the need for the student to achieve in a competitive world and the need to be compassion to his fellow beings [19].

Bina Nusantara University has 6 distinctive core values, namely SPIRIT. SPIRIT should be applied by all the lectures and also the students of Bina Nusantara University, which are nurtured in the active lectures and student. **Striving for excellence** is giving the best not only for ourselves, but where is our market place. Actively participate in the company's vision and mission of Bina Nusantara University. Acting as if the company we work for is our own so we will always give our best. **Perseverance** is diligent in completing responsibilities and giving more than expected. Surely, the basic of our perseverance is our personal responsibility. Never stop trying, let alone give up until it's finished. **Integrity** is the same applies when there is or is not a leader. Many workers only do their jobs well, when there is a leader, Bina Nusantara University urges students to always have an attitude of integrity which means equalizing working time without a leader or with a leader. **Respect** is valuing everyone and work is valuable. Don't look down on or be little anything. From this attitude of respect, tolerance will arise which will open the door of our communication, not closing the possibility of relationships, business partners, etc. **Innovation** is generate new brilliant ideas without erasing existing characteristics. Innovation talks about the present and the future. Innovation means being able to see into the future, knowing what is needed and preparing for what will be needed. **Teamwork** is an application of a combination of all existing soft skills. In a project or group, we cannot think only of ourselves [18].

3 Methods

The research method used in this analysis is linear regression, 150 respondents were involved in this study by distributing questionnaires by random sampling method. All statistical calculations, including regression analysis, were performed using SPSS with a data size of 20KB. The results of the SPSS analysis stored in Ms. Word consist of Descriptive Statistics, Tests of Normality, ANOVA, Coefficients, correlation, and the data stored with a data size of 55KB. The measurement of variables in this study used a Likert scale that was adjusted to the needs of the study. Research variables measured using this scale have a gradation from negative (lowest) to positive (highest), and the target population in this research were students at Bina Nusantara University, Jakarta.

4 Results and Discussion

4.1 Result

Table 1. Descriptive statistics

	N statistic	Range statistic	Min statistic	Max statistic	Mean		SD statistic	Variance statistic
					Statistic	Std. error		
VAR X	150	31.00	114.00	145.00	130.3667	0.60906	7.45944	55.643
VAR Y	150	79.00	131.00	210.00	171.0667	1.53374	18.78442	352.855
Valid N (listwise)	150							

Based on the Table 1 descriptive statistics above, it is known that the minimum score of X variable (Pancasila Education) is 114, the maximum score is 145, the mean is 130.3667, the standard deviation is 7.45944, and the variance is 55.643; meanwhile minimum score Y variable (Value Education) is 131, maximum score is 210, Mean = 171.0667, Standard deviation = 18.78442, and variance = 352.855.

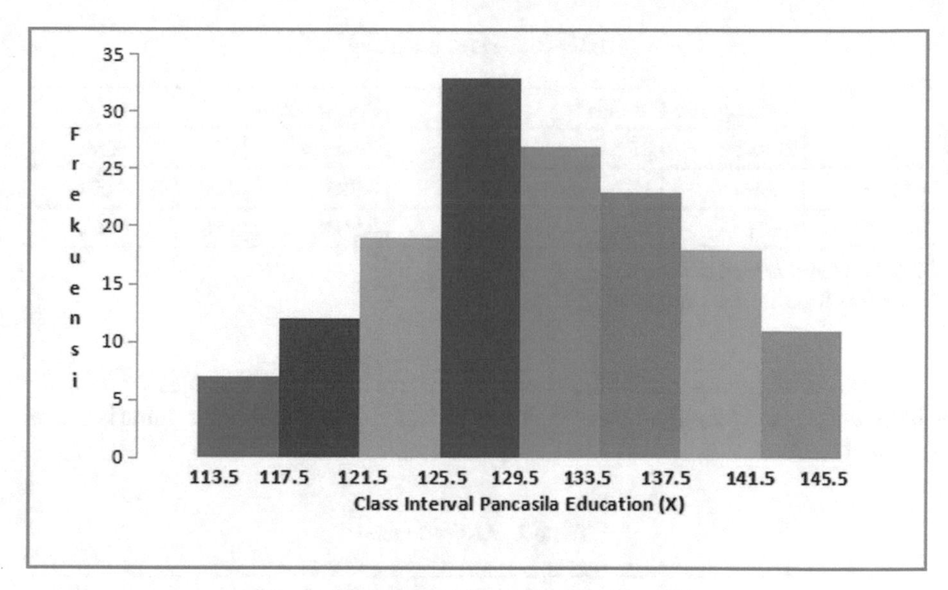

Fig. 1. Histogram of Pancasila Education (X)

From the Fig. 1 histogram of Pancasila Education above showed that the highest class interval distribution is between 126–129.

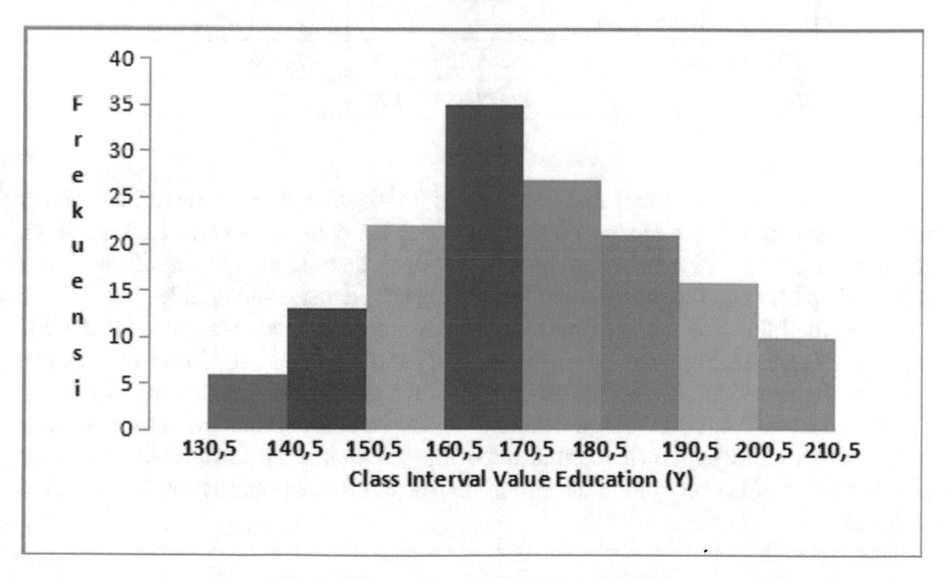

Fig. 2. Histogram of Value Education (Y)

From the Fig. 2 histogram of Value Education above showed that the highest class interval distribution between 161–170.

Table 2. Tests of normality

	Kolmogorov-Smirnov[a]			Shapiro-Wilk		
	Statistic	df	Sig.	Statistic	df	Sig.
VAR X	0.046	150	0.200*	0.985	150	0.091
VAR Y	0.070	150	0.072	0.984	150	0.085

*This is a lower bound of the true significance.
[a]Lilliefors Significance Correction

From the normality test Table 2 above, it is known that all the variables are normally distributed, because the significance value in the Kolmogorov-Smirnov column is greater (>) than 0.050.

Table 3. ANOVA table

			Sum of squares	df	Mean square	F	Sig.
VAR Y * VAR X	Between groups	(Combined)	14307,210	31	461.523	1.423	0.092
		Linearity	4760,404	1	4760.404	14.679	0.000
		Deviation from linearity	9546,806	30	318.227	0.981	0.503
	Within groups		38268,123	118	324.306		
	Total		52575,333	149			

If the value of Deviation from Linearity Sig > 0.05, then there is a significant linear relationship between the independent variable and the dependent variable. Vise versa, if the value of Deviation from Linearity Sig < 0.05, then there is no significant linear relationship between the independent variable and the dependent variable.

From the Table 3 above, the Deviation from Linearity Sig value is obtained. of 0.503 is greater than 0.05. So, it can be concluded that there is a significant linear relationship between the Pancasila Education variable (X) and Value Education variable (Y).

The research focus on the significance of Pancasila education on the value education of the students. The hypothesis formulated in this research is: Pancasila Education has a positive direct effect on Value Education. Statistically the hypothesis can be written as follows:

H_0: $\beta_{yx} \leq 0$

H_1: $\beta_{yx} > 0$

If the significance value (Sig) is smaller (<) than the 0.05 probability, it means that there is an effect of X on Y. On the other hand, if the significance value (Sig) is greater (>) than the probability of 0.05, it means that there is no effect of X on Y.

Table 4. ANOVA[a]

Model		Sum of squares	df	Mean square	F	Sig.
1	Regression	4760.404	1	4760.404	14.735	0.000[b]
	Residual	47814.930	148	323.074		
	Total	52575.333	149			

[a]Dependent Variable: VAR Y
[b]Predictors: (Constant), VAR X

From the Table 4 above, it is known that the significance value (Sig) of 0.000 is smaller than the probability of 0.05, so it can be concluded that H_0 is rejected and H_1 is accepted which means that there is a significant effect of X on Y.

Table 5. Coefficients[a]

Model		Unstandardized coefficients		Standardized coefficients	t	Sig.
		B	Std. error	Beta		
1	(Constant)	72.282	25.776		2.804	0.006
	VAR X	0.758	0.197	0.301	3.839	0.000

[a]Dependent Variable: VAR Y

From the Table 5, it is known that $p_{yx} = 0.301$ and t count (3.839) $> t_{table}$ (1.67) with a significance of 0.000 $< \alpha$ (0.05), then H_0 is accepted and H_1 rejected. So, there is a positive direct effect of Pancasila Education (X) on value education (Y). That is, Pancasila Education influences value education of the students.

4.2 Discussion

More deeply, the test results are known based on the coefficients model, which known the relationship between variables and answers the research hypothesis, directly that Pancasila education has a positive relationship that is 0.301 with value education. And the coefficient regression with p-value < 0,05 means that Pancasila education has significant impact to value education. This result is in line with previous results which show that values education has a significant impact on providing a positive and conducive educational environment, values education also affects institutional culture, increases collaboration between staff and students, and creates a conducive atmosphere for teaching and learning. This factors can improve the attractiveness of students towards the institution [19].

Pancasila Education is education that influences and directs the behavior of students to have awareness of values. Therefore, it is only natural that it contributes significantly to value education to the students. Apparently, Pancasila Education is a variable that influences to value education of the students, so that whatever Pancasila Education is very meaningful for students to have value conscience. This could be because Pancasila education is truly realized by its meaning and role for their awareness of values. It could also because Pancasila education is felt to be truly inspiring for the character building of students. It could also because the lecturer in delivering Pancasila education, really deepens the meaning and significance of the student's awareness of values. It could also because students in the first or second semester were more able to see the meaningfulness of Pancasila Education as values education.

5 Recommendations

The result of this study showed that Pancasila education has significant relationship to value education of the students that present SPIRIT values. In a situation to survive the teaching and learning process in the midst of covid-19, students must have perseverance spirit to produce the best result, honesty as a form of integrity also needs to be built in a pandemic condition, even though learning takes place online but respectful attitude towards lecturers must also be maintained and be guarded. The restrictions on online classes should also not limit the values of teamwork during the student teaching and learning process. Based on the result, it is recommended for the lecturers to keep the Pancasila education to the students, and referring to the pandemic situation, they have to improve more innovative ways to teaching learning process so that the students keep their interest in understanding and implementing the values of Pancasila even though they are still studying in distance learning process. The result of the study will be a recommendation to study program to develop curriculum regarding the value education of Pancasila. For further research, it is suggested to examine other variable such as the intention of the students to value education, and other tools such as smart PLS could be applied for further results to reach more comprehensive result.

6 Conclusions

The Covid-19 pandemic has forced the human community to maintain social distance, and this condition has had a considerable impact on the teaching and learning process of students. In addition to various efforts to deal with climate change and the attack of the covid-19 virus, the sector that must continue to run is education. The most crucial change is the change in the climate of face-to-face learning to online learning which is a challenge for lecturers of Pancasila course. This research is conducted to evaluate the performance character of SPIRIT through Pancasila education during the Covid-19. The study showed that Pancasila education has significant relationship to value education of the students that present SPIRIT values. The results of the study can be a recommendation to lecturer to explore many variative ways to teach Pancasila education. Next research can be conducted with more size of sample and more references because there are not many sources that research specifically about SPIRIT value education.

The result of this study showed that Pancasila education has significant relationship to value education of the students that present SPIRIT values. In a situation to survive the teaching and learning process in the midst of covid, students must have perseverance spirit to produce the best result, honesty as a form of integrity also needs to be built in a pandemic condition, even though learning takes place online but respectful attitude towards lecturers must also be maintained and be guarded. The restrictions on online classes should also not limit the values of teamwork during the student teaching and learning process. Based on the result, it is recommended for the lecturers to keep the Pancasila education to the students, and referring to the pandemic situation, they have to improve more innovative ways to teaching learning process so that the students keep their interest in understanding and implementing the values of Pancasila even though they are still studying in distance learning process. The result of the study will be a recommendation to study program to develop curriculum regarding the value education of Pancasila. For further research, it is suggested to examine other variable such as the intention of the students to value education, and other tools such as smart PLS could be applied for further results to reach more comprehensive result.

References

1. Zang, S.M., Benjenk, I., Breakey, S., Pusey-Reid, E., Nicholas, P.K.: The intersection of climate change with the era of COVID-19. Public Health Nurs. **38**(2), 321–335 (2021). https://doi.org/10.1111/phn.12866
2. Upoalkpajor, J.-L.N., Upoalkpajor, C.B.: The impact of COVID-19 on education in Ghana. Asian J. Educ. Soc. Stud. **9**(1), 23–33 (2020). https://doi.org/10.9734/ajess/2020/v9i130238.)
3. Jena, K.P.: Research article impact of pandemic Covid-19 Covid. Int. J. Curr. Res. **12**(7), 12582–12586 (2020). http://journalcra.com/article/impact-pandemic-covid-19-education-india
4. Silalahi, R., Yuwono, U.: Research in social sciences and technology RESSAT. Res. Soc. Sci. Technol. **3**(1), 109–121 (2018). https://www.learntechlib.org/p/187543/
5. Katılmış, A.: Values education as perceived by social studies teachers in objective and practice dimensions. Kuram ve Uygulamada Egit. Bilim. **17**(4), 1231–1254 (2017). https://doi.org/10.12738/estp.2017.4.0570
6. Kristiono, N.: Penguatan Ideologi Pancasila di Kalangan Mahasiswa Universitas Negeri Semarang. Harmony **2**(2), 193–204. (2017). https://journal.unnes.ac.id/sju/index.php/harmony/article/view/20171/9563
7. Sitorus, J.H.E.: Pancasila-based social responsibility accounting. Proc. - Soc. Behav. Sci. **219**, 700–709 (2016). https://doi.org/10.1016/j.sbspro.2016.05.054
8. Aneja, N.: The importance of value education in the present education system & role of teacher. Int. J. Soc. Sci. Humanit. Res. **2**(3), 230–233 (2014). www.researchpublish.com
9. Duban, N., Aydoğdu, B.: Values education from perspectives of classroom teachers. Eur. J. Soc. Sci. Educ. Res. **7**(1), 80 (2016). https://doi.org/10.26417/ejser.v7i1.p80-88
10. Vijaya Lakshmi, V., et al.: Value education in educational institutions and role of teachers in promoting the concept. Int. J. Educ. Sci. Res. **8**(4), 29–38 (2018). https://doi.org/10.24247/ijesraug20185
11. Anggraini, D., Fathari, F., Anggara, J.W., Ardi Al Amin, M.D.: The practice of Pancasila values for the millennial generation. J. Inov. Ilmu Sos. dan Polit. **2**(1), 11 (2020). https://doi.org/10.33474/jisop.v2i1.4945
12. Amir, S.: Pancasila as integration philosophy of education and national character. Int. J. Sci. Technol. Res. **2**(1), 54–57 (2013). www.ijstr.org
13. Jannah, F., Fahlevi, R.: Strengthening the Pancasila character values in forming the character of Pancasilais generation. In: Proceedings of the 1st International Conference on Creativity,

Innovation and Technology in Education (IC-CITE 2018), vol. 274, no. 18, pp. 77–80 (2018). https://doi.org/10.2991/iccite-18.2018.18
14. Krisnamukti, D.B.P.: Implementation of Pancasila values in the lives of students of the Faculty of Agriculture, Universitas Brawijaya in Malang. J. Rontal Keilmuan PKn **6**(1), 66–72 (2020)
15. Nurgiansah, T.H.: Pancasila education as an effort to build honest character. Jurnal Pendidikan Kewarganegaraan Undiksha **9**(1), 33–41 (2021). https://ejournal.undiksha.ac.id/index.php/JJPP
16. Şahin, Ü.: Values and values education as perceived by primary school teacher candidates. Int. J. Progress. Educ. **15**(3), 74–90 (2019). https://doi.org/10.29329/ijpe.2019.193.6
17. Indrani, B.: Importance of value education in modern time. Educ. India J. A Q. Ref. J. Dialog. Educ. **1**(August), 32 (2012)
18. Benawa, A., Lake, S.C.J.M.: The effect of social media and peer group on the spirit characters formation of the students. Int. J. Eng. Adv. Technol. **8**(6) (Special Issue 3) (2019). https://doi.org/10.35940/ijeat.F1062.0986S319
19. Singh, A.: Evaluating the impacts of value education: some case studies. Int. J. Educ. Plan. Adm. **1**(1), 1–8 (2011). http://www.ripublication.com/ijepa.htm

Is It Possible to Instilling Character SPIRIT Values Through Civic Education During COVID-19 Pandemic?

Alfensius Alwino[1(✉)], Arcadius Benawa[2], Iwan Irawan[1], Dian Anggraini Kusumajati[3], and Erma Lusia[4]

[1] Character Building Development Center, Computer Science Department, School of Computer Science, Bina Nusantara University, West Jakarta, Indonesia
{alfensius.alwino,iwan.irawan}@binus.ac.id

[2] Character Building Development Center, Mass Communication Department, Faculty of Economics and Communication, Bina Nusantara University, West Jakarta, Indonesia
aribenawa@binus.ac.id

[3] Character Building Development Center, Information System Department, School of Information System, Bina Nusantara University, West Jakarta, Indonesia
diananggi@binus.ac.id

[4] Tourism Department, Faculty of Economics and Communication, Bina Nusantara University, West Jakarta, Indonesia
ermalusia@binus.ac.id

Abstract. When all nations including Indonesia are hit by the COVID-19 pandemic, we are compelled to conduct research with the purpose is to see students' understanding of Civic Education towards Value Education of the Students. The method of this research used the quantitative method by distributing questionnaires to 150 respondents as simple random sampling and then data was processed by the SPSS 22 program and completed by using path analysis. The results showed that Civic Education has no significant influence, which is indicated by the significance value in the ANOVA table and the coefficient table of 0.164 which means it is bigger than the 0.05 probability. The result showed that Civic education has no significant influence on the Value Education of the students. The implication is that Civic Education should be delivered more intensively and seriously so that instilling values to the students as the young generation through Civic Education is more effective. By that, the students are not only having an understanding of Civic Education but also get to implement the values even though statistically there is no significant effect.

Keywords: Civic education · Value education · COVID-19 pandemic

1 Introduction

The COVID-19 pandemic has affected the student teaching and learning process, including the number of closed classes and the shifting of the learning process to online [1]. Even so, the lockdown still provides opportunities for lectures and students to continue

© The Author(s), under exclusive license to Springer Nature Switzerland AG 2023
L. C. Tang and H. Wang (Eds.): BDET 2022, LNDECT 150, pp. 117–126, 2023.
https://doi.org/10.1007/978-3-031-17548-0_11

to interact, teachers give assignments to students via the internet, lectures deliver it through live conferences using various applications such as Zoom, Google meet, Facebook, Youtube, and Skype etc. [2]. This is certainly a challenge for teachers in adjusting the way of learning for students.

Meanwhile Character education in Indonesia aims to build the potential of students to become human beings who have the skills, desires, and good morals as stated in the Law on the National Education System. [3]. Education prepares students to become individuals who have knowledge, skills, attitudes, values, and behaviors that will have a positive impact on students' daily lives in society. [4]. From Civic Education according to Mali Benyamin [5], there are many achievements that have been achieved by Indonesia as a sovereign country. Among them are increasingly democratic elections, law enforcement and human rights (HAM), better citizen welfare, more advanced education, health services, more equitable infrastructure, etc. Nevertheless, there are several challenges and obstacles that can undermine the vision and mission of the Indonesian state to truly become an independent, sovereign, just and prosperous country. Among them are rampant corruption, forest destruction, moral decline, religious fundamentalism, terrorism, capitalism, identity politics, political oligarchy, discrimination, weakening of Pancasila ideology, hoaxes, fake news, etc. At the student level, the problems that often arise are bullying, hoaxes, cheating, plagiarism, racism, apathy with the condition of the nation, and less concerned with threats to state ideology. If these challenges and obstacles are not overcome, Indonesia could fall into a collapse state.

The conditions needed by the Indonesian state to truly become an independent, sovereign, just and prosperous country are honesty, justice, kindness, hard work, cooperation (mutual assistance), law enforcement and human rights, transparent and accountable state management, healthy political competition, citizens' commitment to maintain the state ideology, and upholding multiculturalism. Such values are not automatically embedded in their citizens. The installation of such values can be done through an institutionalized educational process. Higher Education is the most strategic place to instill these values in a systematic and institutionalized manner.

Law Number 12 of 2012 concerning the state requires every university to systematically and institutionally teach Civic Education in order to form the students who have value such as honest, fair, kind, love the homeland, patriotic, innovative, responsible, have integrity, and are respectful [6]. The state is aware that its can be formed through the educational process. Therefore, the state makes Civic Education a stand-alone subject and must be studied by every student.

However, the Covid-19 pandemic has had an impact on the process of inserting value education by civic education. The formation of values, such as honest, fair, kind, patriotic, innovative, integrity citizens, etc. is disrupted. The same disruption when students internalize the SPIRIT value which is the specialty of Bina Nusantara University. At Bina Nusantara University, every student needs to establish himself with SPIRIT values. SPIRIT: Striving for excellence, Perseverance, Integrity, Respect, and Innovation. SPIRIT values are very important for students, because they are the values that shape a person's character. In the book *Full Steam Ahead! Unleash the Power of Vision in Your Work and Your Life, 2nd Edition*, Ken Blanchard, Jesse Lyn Stoner, explained that values are deep beliefs that shape a person's character [7].

The elaboration of the SPIRIT Values includes Striving for excellence, which is revealed in a diligent and tough attitude, Perseverance is revealed in the value of discipline and not giving up, Integrity is revealed in honest, transparent behavior, dare to do what is right. In addition, the value of Respect is revealed in fair, non-discriminatory behavior, Innovation is revealed in the behavior of always looking ahead, giving birth to brilliant new ideas, and finally Teamwork which is revealed in the value of collaboration and mutual building [8].

Both students and educators are more focused on efforts to save themselves and others, so they don't get infected with Covid-19. Or if they are infected with Covid-19, educators and students focus on the recovery process. At the same time, the character of caring towards oneself and others grows stronger.

2 Methods

This study used quantitative methods, namely examining the population using research measuring instruments, which continues with quantitative data analysis with the aim of testing assumptions or hypotheses. The quantitative method in this study was conducted through a survey, namely distributing questionnaires to 150 respondents. Questionnaires were distributed to 150 students at Bina Nusantara University. Respondents were selected using the Simple Random Sampling technique. In the book Introduction to Research in Education Donald Ari, explained survey analysis to test the established hypotheses. For the method of calculating the sample in this study, it was carried out using the Slovin formula. [9].

The Slovin formula is the formula used to calculate the minimum number of samples in a finite population survey. The main purpose of the survey is to estimate the proportion of the population. What is estimated is the proportion of the population, not the population mean. The value of the margin of error is determined by the researcher himself. The smaller the size of the error that is desired or set, then of course the larger the sample size obtained. The data were processed using the SPSS 22 program, then completed using path analysis. SPSS (Statistical Program for Social Science) is a computer application program package for analyzing statistical data. With SPSS we can create reports in the form of tabulations, charts (graphs), plots (diagrams) of various distributions, descriptive statistics, and complex statistical analyses.

3 Literature Review

Civic Education is a science that studies the position and role of citizens in carrying out their rights and obligations according to and within the limits of constitutional and legal provisions. Nurhasanah & Saputra defined civic education as a basic subject in schools, which is designed to prepare young citizens to take an active role in society in the future [10]. Civic Education has a clear legal basis, namely Law Number 12 of 2012 [6].

In the vision and mission of Civic Education for Higher Education, it is explicitly stated about values such as morals, responsibility, and love for the homeland. The vision of Civic Education is as a source of values and guidelines in the development and implementation of study programs, in order to help in strengthening their personalities as whole people. While the mission of Civic Education is to help students strengthen their personality so that they are consistently able to realize the values of Pancasila, a sense of nationality, love for the homeland, and are able to develop science and technology responsibly and morally. The competencies expected of a student after studying Civic Education are to become a professional scientist who is democratic, civilized, and has a sense of love for the nation and state, as well as being honest, fair, good, innovative, competitive, disciplined, participating, and critical citizens in the nation and state.

According to Budi Utomo, Civic Education is a process of realization to be an Indonesian, which is a systematic and planned effort from the State to make Indonesians have an Indonesian character [11]. Civic Education is a democratic education that can educate the younger generation to become democratic citizens. Civic Education is political education that aims to help students become politically mature citizens and participate in building a democratic political system. That is, Civic Education is an instrument to strengthen democracy itself. In essence, Civic Education is an educational program that balances the cognitive and character dimensions, which includes nationality, citizenship, democracy, human rights and civil society, as well as applying the principles of democracy and humanistic education [12].

Civic Education does not only emphasize knowledge about citizenship, but more than that to form students with character and personality. Civic Education is a very basic education for students because it instills morals in them from an early age [13]. Even civic education is one of the subjects that becomes the leading sector in developing the character of students [14].

Endang Komara stated that the content of Civic Education does not necessarily shape the character of students [15]. The skills of the teacher or lecturer are still needed to develop an interesting, fun, challenging learning process, and shape students to be able to think critically and constructively. So, teachers or lecturers must be able to present contextual learning materials, linking the subject matter with real conditions in the field. Between theory and practice must be in line. In addition, students should also be encouraged to be able to identify problems and encourage them to solve the problems themselves.

Value education is a teaching method that aims to help students understand, realize and experience values and students are able to place them integrally in their daily lives. Here we can draw a conclusion that the orientation of value education is to guide the conscience so that humans develop more positively gradually and continuously in order to develop for the better and positively, besides that it also aims to be able to instill good values into the human person and keep bad values away. What is expected here is that students are able to experience the process of transformation and trans-internalization of values.

Value education has the same orientation as morality education, both of which are committed to what steps students must take in order to have values and virtues that will shape them into good human beings. Values and norms that enable them to be able to make responsible decisions about their life's problems.

As a series of the identity of a nation, character is the basic value of behavior that becomes a reference for the values of interaction between humans. Universally, various characters are formulated as values of living together based on pillars, peace, respect, cooperation, freedom, happiness, honesty, humility, love, responsibility, simplicity, tolerance and unity.

Value education plays a role in assessing the criteria for values consisting of awareness, emotion, excitement and integration of individual's behavior, and the main purpose of values education is to make values permanent behaviors in students' daily. Value education prepares the need for the student to achieve in a competitive world and the need to be compassion to his fellow beings [16]. Value education plays a role in assessing the criteria for values consisting of awareness, emotion, excitement and integration of individual's behavior, and the main purpose of value education is to instill good behavior so that it becomes permanent behavior in students' daily lives. Value education prepares the need for the student to achieve in a competitive world and the need to be compassion to his fellow beings.

Bina Nusantara University has 6 distinctive core values, namely SPIRIT. SPIRIT should be applied by all the lectures and also the students of Bina Nusantara University, which are nurtured in the active lectures and student. Striving for excellence is giving the best not only for ourselves, but where is our marketplace. Actively participate in the company's vision and mission of Bina Nusantara University. Acting as if the company we work for is our own so we will always give our best. Perseverance is diligent in completing responsibilities and giving more than expected. Surely, the basic of our perseverance is our personal responsibility. Never stop trying, let alone give up until it's finished. Integrity is the same applies when there is or is not a leader. Many workers only do their jobs well, when there is a leader, Bina Nusantara University urges students to always have an attitude of integrity which means equalizing working time without a leader or with a leader. Respect is valuing everyone, and work is value able. Don't look down on or be little anything. From this attitude of respect, tolerance will arise which will open the door of our communication, not closing the possibility of relationships, business partners, etc. Innovation is generated new brilliant ideas without erasing existing characteristics. Innovation talks about the present and the future. Innovation means being able to see into the future, knowing what is needed and preparing for what will be needed. Teamwork is an application of a combination of all existing soft skills. In a project or group, we cannot think only of ourselves [17] & [18].

In short, the contribution of civic education to the installation of values, as stated by the researchers above, can be seen in the following table (Table 2):

Table 1. The contribution of civic education to the installation of values

Researcher	Types of Values Civic Education Contribution Results
Nurhasanah & Saputra	• Civic education designed to prepare young citizens to take an active role in society in the future • Civic education can improve morality • Strengthen students' personality • Responsibility • Love for the homeland • Become a professional scientist who is democratic, civilized, and has a sense of love for the nation and state, as well as being honest, fair, good, innovative, competitive, disciplined, participating, and critical citizens in the nation and state
Budiutomo	• To help students become politically mature citizens and participate in building a democratic political system • Civic Education is an instrument to strengthen democracy itself • Civic Education is an educational program that balances the cognitive and character dimensions, which includes nationality, citizenship, democracy, human rights, and civil society, as well as applying the principles of democracy and humanistic education. So Civic education is a process "Indonesianisasi" (to make Indonesians have an Indonesian)
Endang Komara	• To help students understand, realize and experience values and students can place them integrally in their daily lives • To guide the conscience so that humans develop more positively gradually and continuously in order to develop for the better and positively • To be able to instill good values into the human person

4 Results and Discussion

4.1 Result

Table 2. Descriptive statistics

	N	Range	Min	Max	Mean		SD	Variance
	Statistic	Statistic	Statistic	Statistic	Statistic	Std. Error	Statistic	Statistic
VAR-X	150	31.00	129.00	160.00	144.7200	0.63627	7.79272	60.726
VAR-Y	150	79.00	131.00	210.00	171.0667	1.53374	18.78442	352.855
Valid N (listwise)	150							

From the Table 1 descriptive statistics, it is known that the minimum score of X (Civic Education) is 129, maximum score is 160, Mean = 144.7200, Standard deviation = 7.79272, and variance = 60.726; mean while minimum score Y (Value Education)

is 131, maximum score is 210, Mean = 171.0667, Standard deviation = 18.78442, and variance = 352.855.

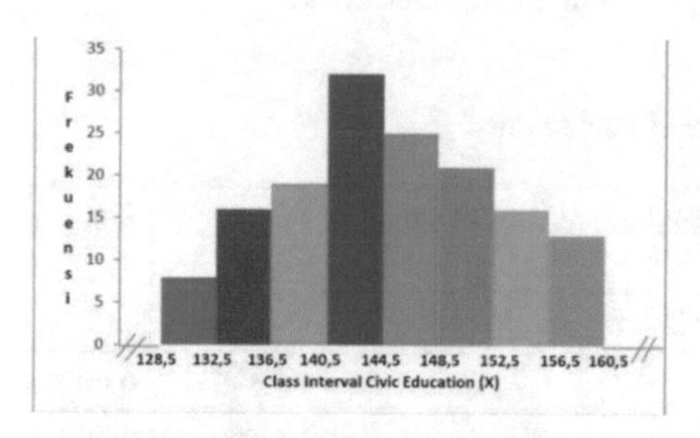

Fig. 1. Histogram of civic education

Fig. 2. Histogram of value education

From Fig. 1 histogram of Civic Education, it was known that the class interval highest distribution is between 141–144.

From Fig. 2 histogram of the Value Education, it was known that the class interval highest distribution is between 161–170.

Table 3. Tests of normality

	Kolmogorov-Smirnov[a]			Shapiro-Wilk		
	Statistic	Df	Sig.	Statistic	df	Sig.
VAR-X	0.061	150	0.200*	0.983	150	0.064
VAR-Y	0.070	150	0.072	0.984	150	0.085

*This is a lower bound of the true significance.
a. Lilliefors Significance Correction

From the Table 3 test of normality above, it is known that all of the variables are normally distributed, because the significance value in the Kolmogorov-Smirnov column is greater (>) than 0.050.

Results of Significance Test

Table 4. Table of ANOVA[a]

Model	Sum of squares	df	Mean square	F	Sig.
1 Regression	687.320	1	687.320	1.960	0.164[b]
Residual	51888.013	148	350.595		
Total	52575.333	149			

a. Dependent Variable: VAR-Y.
b. Predictors: (Constant), VAR-X.

Based on the Table 4 above, it is known that the significance value (Sig) of 0.164 is greater than the probability of 0.05, so it can be concluded that H0 is accepted and H1 is rejected, which means that there is no significant effect of X on Y.

Table 5. Table of coefficients[a]

Model	Unstandardized coefficients		Standardized coefficients	t	Sig.
	B	Std. error	Beta		
(Constant)	131.180	28.528		4.598	0.000
VAR-X	0.276	0.197	0.114	1.400	0.164

a. Dependent Variable: VAR-Y.

If the significance value (Sig) is smaller (<) than the probability of 0.05, it means that there is an effect of X on Y; on the other hand, if the significance value (Sig) is greater (>) than the probability of 0.05, it means that there is no effect of X on Y.

Based on the Table 5 above, it is known that the significance value (Sig) of 0.164 is greater than the probability of 0.05, so it can be concluded that H0 is accepted and H1 is rejected, which means that there is no significant effect of X on Y.

4.2 Discussion

The last few years have shown the lack of student interest in Pancasila, Civic, and Religion Courses [19]. Other researcher observed that the attitudes seen from some people and students showed attitudes that not applied Pancasila, Civic, and Religion values in everyday life. Therefore, it is very necessary for Pancasila, Civic, and Religion education to be taught at the school and college level [20].

Various factors that cause the decline in knowledge and understanding of Pancasila, Civic, and Religion among students must be explored and the best solution sought to re-strengthen knowledge and understanding of Pancasila, Civic, and Religion among students. Based on that phenomenon, the research aims to reveal the significance of Pancasila, Civic, and Religion education on the value education of the students, especially in Bina Nusantara University, which well known as SPIRIT values.

The results of this study indicated that Value Education, including Core Values at Bina Nusantara, is very complex which influences it. It cannot only be seen from one, two or even three courses which are better known as general basic courses which in the era of the Minister of Education, Nadiem Makariem who are now proclaimed as compulsory curriculum subjects. That is, in terms of formal education alone, it is not enough to instill values in general subjects such as Civic Education. The findings of this study convince us that all elements of formal and non-formal education must contribute to inculcating values in students at every level, not least at the tertiary level.

5 Conclusions

Covid-19 pandemic has forced mankind to maintain social distance. This condition has had a considerable impact on the teaching and learning process of the students. In addition to various efforts to deal with climate change and exacerbated by the Covid-19 virus attack, education must, however, continue to be carried out. Inevitably, the most crucial change is the change in the learning climate: from face-to-face learning to online learning which is a challenge for civics course lecturers. And from this research we concluded that Civic Education has an important role in the context of civic values education, but in the perspective of values education in general, Civic Education did not have a significant influence. So, answering the question posed in the title of this article: Is it possible to instil values through Civic Education in the COVID-19 pandemic can be answered that it is possible for values related with citizenship, but it is not possible to instil values in general just only through civic education.

That is why, especially during the COVID-19 pandemic, the inculcation of universal values cannot only rely on civic education, because civic education only targets the cultivation of civic values. Instilling other values can be through family, community, and other institutions.

Of course, citizenship values do not stop being installed through education, but what is more important in the future is that there must be a process of habituation or habituation to live the values of citizenship.

References

1. Upoalkpajor, J.L.N., Upoalkpajor, C.B.: The impact of COVID-19 on Education in Ghana. Asian J. Educ. Soc. Stud. **9**(1), 23–33 (2020). https://doi.org/10.9734/ajess/2020/v9i130238.)
2. Jena, K.P.: Research article impact of pandemic Covid-19 Covid. Int. J. Curr. Res. **12**(7), 12582–12586 (2020)
3. Silalahi, R., Yuwono, U.: Research in social sciences and technology ressat. Res. Soc. Sci. Technol. **3**(1), 109–121 (2018)
4. Katılmış, A.: Values education as perceived by social studies teachers in objective and practice dimensions. Kuram ve Uygulamada Egit. Bilim. **17**(4), 1231–1254 (2017)
5. Mikhael, M.B., et al.: Pendidikan Kewarganegaraan. Upaya Mengindonesiakan Orang Indonesia. Immaculata Press, Bekasi (2016)
6. Undang-Undang Nomor 12 Tahun 2012 & SK Dirjen Dikti No.43/DIKTI/KEP/2006
7. Blanchard, K., Stoner, J.L.: Full Steam Ahead! Unleash the Power of Vision in Your Work and Your Life, 2nd edn., pp. 59–68. Berrett-Koehler Publishers, Oakland (2011)
8. Vison & Mission. https://binus.ac.id/vision-mission. Accessed 21 Mar 2022
9. Donald, A.: Introduction to Research in Education. Wadsworth Publishing, Belmont (2009)
10. Nurhasanah, N., Saputra, M.A.K.: Pengaruh Penggunaan Metode Problem Solving terhadap Hasil Belajar PKN tentang Globalisasi di Kelas IV Sekolah Dasar di Desa Sukaharja Kecamatan Cijeruk Bogor. J. Ilmiah PGSD **10**(2), 77–85 (2016)
11. Budiutomo, T.: Pendidikan Kewarganegaraan dalam Membentuk Karakter Bangsa. Acad. Educ. J. Pendidikan Pancasila dan Kewarganegaraan **4**(1), 32–38 (2013)
12. Dewi, R.R., Suresman, E., Suabuana, C.: Pendidikan Kewarganegaraan Sebagai Pendidikan Karakter di Persekolahan. ASANKA: J. Soc. Sci. Educ. **1**(2), 71–84 (2021)

13. Narimo, S., Sutama, S., Novitasari, M.: Pembentukan Karakter Peserta Didik dalam Pembelajaran Pendidikan Pancasila dan Kewarganegaraan Berbasis Budaya Lokal. J. Varia Pendidikan **31**(1), 39–44 (2019)
14. Dianti, P.: Integrasi Pendidikan Karakter dalam Pembelajaran Pendidikan Kewarganegaraan untuk Mengembangkan Karakter Siswa. J. Pendidikan Ilmu Sosial **23**(1), 58–68 (2014)
15. Komara, E.: Curriculum and civic education teaching in Indonesia. EDUCARE: Int. J. Educ. Stud. **10**(1), 23–32 (2017)
16. Indrani, B.: Importance of value education in modern time. Educ India J.: A Q. Refereed J. Dialogues Educ. **1**(3), 35–42 (2012)
17. Benawa, A., Martoredjo, N.T., Laurentius, L.Y.: The effect of lecturer leadership and organizational culture on the students' SPIRIT character building. In: 10th International Conference on Human System Interactions (HSI), pp. 116–118 (2017)
18. Benawa, A., Silverius Lake, C.J.M.: The effect of social media and peer group on the spirit characters formation of the students. Int. J. Eng. Adv. Technol. 8(6 Special Issue 3) (2019)
19. Kristiono, N.: Penguatan Ideologi Pancasila di Kalangan Mahasiswa Universitas Negeri Semarang. Harmony **2**(2), 193–204 (2017)
20. Anggraini, D., Fathari, F., Anggara, J.W., Al Amin, M.D.: Pengamalan nilai-nilai Pancasila bagi generasi milenial. J. Inovasi Ilmu Sosial Dan Politik (JISoP) 2(1), 11–18 (2020)

An Adaptive Self-detection and Self-classification Approach Using Matrix Eigenvector Trajectory

Chuan Jiang[1(✉)] and Li Chen[2]

[1] Tech Center, CRRC Qingdao Sifang Rolling Stock Research Institute Co., Ltd., Shanghai 200333, China
jiangcn@mail.uc.edu
[2] International School of Photoelectronic Engineering, Qilu University of Technology, Jinan 250353, China

Abstract. Motivated by the rapid development of the next generation artificial intelligence, we propose a novel adaptive self-detection and self-classification algorithm using matrix eigenvector trajectory in the paper. This algorithm's mathematical inferences are also described and proved theoretically. The proposed algorithm is used in a multi-class bearing faults classification problem to validate its effectiveness. Results show that in an online data processing scenario, it can automatically adapt to new data patterns so that self-detection and self-classification can be realized by monitoring the eigenvector evolution trajectory. By comparing with other machine learning algorithms, we have validated that the proposed algorithm does not require explicit training, its required data processing time dropped more than 78% and achieved the same classification accuracy on new testing data.

Keywords: Self-detection · Self-classification · Eigenvector trajectory · Adaptability · Self-learning

1 Introduction

The world we live in has been changed tremendously with the development of Internet of Things (IoT) and 5G network [1, 2]. From the way we drive to how we make purchases, smart sensors and chips are embedded in the physical things that surround us, each transmitting large amount of data. Data that let us better understand how these things work and work together [3]. But how do we specifically make the most sense of the IoT data and turn them into actionable information? The knowledge of artificial intelligence, particularly, machine learning is the key [4]. To meet the challenges and requirements in today's sophisticated systems, faster, smarter, and more automated algorithms need to be developed to overcome the limits of the legacy machine learning. As pointed in [5], the next generation artificial intelligence and machine learning must be intelligent, self-learning and adaptive. Adaptability and self-learning are two major identified capacities that new algorithms should be improved on [6]. In order to automatically detect new

© The Author(s), under exclusive license to Springer Nature Switzerland AG 2023
L. C. Tang and H. Wang (Eds.): BDET 2022, LNDECT 150, pp. 127–137, 2023.
https://doi.org/10.1007/978-3-031-17548-0_12

abnormality, they must be able to update their model parameters adaptively to thrive in new environments, learn from each individual cases, respond to various situations in different ways. They are also expected to track, adapt to the specific situation of every entity of interest over time, and provide real time actionable insights continuously.

In this paper, we propose a novel adaptive self-detection and self-classification algorithm using matrix eigenvector trajectory. In an online data processing mode, this algorithm can monitor the eigenvector trajectory evolution and use its direction as a key indicator for self-detection and self-classification. This algorithm does not require explicit training using large amount of historical data. It is light-weight, fast, and easy to implement as a batch or an adaptive technique. In addition, this algorithm is not domain specific. It is potentially applicable to various types of data, such as cybersecurity data, mobile data, business data, social media data, health data, etc.

The remainder of this paper is organized as follows. In Sect. 2, the existing artificial intelligence and machine learning technologies are reviewed. Their applications, benefits and shortcomings are also discussed. In Sect. 3, we introduce the mathematical procedure and prove its inferences. The proposed algorithm is applied to a multi-fault bearing classification to demonstrate its effectiveness in Sect. 4. Section 5 draws the conclusion.

2 State-of-the-Art

Numerous efforts have been targeted on anomaly detection and classification using artificial intelligence and they could be categorized into two types. (1) Machine learning: Support Vector Machine (SVM) was used to discriminate the stages of sleep which can help physicians to do an analysis and examination of related sleep disorders [7]. The authors in [8] applied Decision Tree (DT) algorithm to formulate a model for improving the culvert inspection process. As an unsupervised algorithm, the Self-Organizing Map (SOM) demonstrated its feasibility in personalized mental stress detection both in constrained and free-living environment [9]. XGBoost [10], Statistical Pattern Recognition (SPR) [11], and Fuzzy Clustering [12], also proved to be effective in stock market volatility analysis, structural health monitoring, and COVID-19 spread rate comparison, respectively. (2) Deep learning: Convolutional neural networks (CNN) helped extract the fusion feature by jointing the spatial and spectral information simultaneously and achieve good image classification accuracy [13]. In [14], CNN was also used for establishing intelligent spectroscopic models for evaluating the level of water pollution, and provide smart technical support in dealing with the issues of water recycling and conservation for agricultural cultivation. Authors in [15] developed a Recurrent Neural Network (RNN) based Long Short-term Memory (LSTM) called RNN-LSTM to realize predictive and proactive maintenance for high speed railway power equipment. However, most of current machine learning and deep learning algorithms have obvious limitations. For instance, both extract knowledge limited to the specific set of historical data, and information can only be obtained and interpreted with regards to previous patterns learned from the data. Not only they require explicit and relatively time-consuming training with vast amount of data, but also they terminate training when the condition of an objective function is met or pre-defined convergence is reached. This training mechanism does not allow automatic model parameter updates with new incoming data so that they don't have the ability to learn from new patterns continuously. In addition, complicated models like deep learning algorithms lack interpretability as they are not able to explain their decision-making. This may limit its effectiveness in many industries and fields. Last, most of these algorithms can only perform detection or classification at one time. If both tasks are required, chances are another model needs to be trained separately.

To accommodate these issues, this paper proposes a novel adaptive self-detection and self-classification algorithm using matrix eigenvector trajectory. This algorithm is light-weight, easy to use, and can adapt to new patterns hidden in new incoming data in an online mode. Moreover, by monitoring the eigenvector trajectory evolution of feature matrix, this algorithm can potentially complete self-detection and self-classification simultaneously when both baseline and abnormal data are provided.

3 Proposed Algorithm

3.1 Algorithm Procedure

In this section we describe a new algorithm that enables self-detection and self-classification in an online data processing mode. Furthermore, this algorithm's mathematical inferences are provided to demonstrate its effectiveness theoretically.

Specifically, we revised the method proposed in [16] and extended it to make it suitable for an online self-detection and self-classification scenario. In more detail, the procedure is designed as follows:

1) For an n-dimensional dataset X which consists of multiple classes. Use x_i to represent the *i-th* sample in X. For simplicity, we assume X has two classes and there are s and t samples in each class, respectively:
   Class 1: $x_i = (d_{i1}, d_{i2}, \ldots, d_{in}), i = 1, 2, \ldots, s$
   Class 2: $x_i = (d_{i1}, d_{i2}, \ldots, d_{in}), i = s + 1, s + 2, \ldots, s + t.$
2) Instead of subtracting the mean of X, we only subtract the mean of Class 1 (baseline) for X. In other words, the mean vector is not calculated from all columns in X, it is calculated from all columns of the first s rows in X instead.
3) Initialize an n-dimensional eigenvector e_k^0 with an extremely small magnitude, $k = 1, 2, \ldots, n$.
4) Calculate the *k-th* eigenvector e_k based on the following equations. The purpose of Eq. (3.1) is to map the *i-th* sample to the latest *k-th* principal component vector. So y_i is the projection on the direction of *k-th* principal component vector. To calculate the change of the *k-th* principal component vector, a Hebbian term is defined in Eq. (3.2). Then in Eq. (3.3), the latest *k-th* eigenvector can be obtained by normalization. The *(i + 1)-th* sample will be used to update the *k-th* eigenvector to e_k^{i+1}.

$$y_i = \frac{1}{\left\| e_k^{i-1} \right\|} \left(e_k^{i-1} \right)^T x_i \tag{3.1}$$

$$\phi(y_i, x_i) = y_i x_i \tag{3.2}$$

$$e_k^i = \frac{\sum_i \phi(y_i, x_i)}{\left\| \sum_i \phi(y_i, x_i) \right\|} \tag{3.3}$$

5) Up to this point, the *k-th* eigenvector has been obtained after using all samples. To find the *(k + 1)-th* eigenvector, the following equation must be used for all data samples to remove the effect of the *k-th* eigenvector to avoid being found again:

$$x_i' = x_i - \left(e_k^T x_i\right) e_k \tag{3.4}$$

6) Repeat Step 3)–5) until all n eigenvectors have been obtained. Alternatively, it could be repeated m times as long as $m \leq n$.

3.2 Mathematical Inference

Based on the mathematical steps of this algorithm, we can infer that:

a) The trajectory evolution of e^i will be moving towards Class 2. It will be more linear if Class 2 is more aggregated.
b) $\|e^i\| > \|e^{i-1}\|$, $i = s + 1, s + 2, \ldots, s + t$, before normalization.

The mathematical proof is provided as follows:

Proof: after Step 2) in Sect. 3.1, the data in baseline Class 1 is already centered around O in the n-dimensional coordinate.

$x_i = (d_{i1}, d_{i2}, \ldots, d_{in})$, $i = 1, 2, \ldots, s$

There exists a small positive value ε so that:

$$|d_{ij}| < \varepsilon,\ i = 1, 2, \ldots, s \text{ and } j = 1, 2, \ldots, n \tag{3.5}$$

Since we have:

$$e^i = \sum\nolimits_i \phi(y_i, x_i) = y_i x_i + e^{i-1} = \sum\nolimits_i y_i x_i = y_i x_i + y_{i-1} x_{i-1} + \ldots + y_1 x_1 \tag{3.6}$$

$$\|x_i\| = \sqrt{d_{i1}^2 + d_{i2}^2 + \ldots + d_{in}^2} < \sqrt{n\varepsilon^2} = \varepsilon\sqrt{n} \tag{3.7}$$

So, it can be obtained that:

$$\left\|e^i\right\| \leq \sum\nolimits_i |y_i| \|x_i\| < \varepsilon\sqrt{n} \sum\nolimits_i |y_i| \tag{3.8}$$

Given ε is a small positive value, e^s can be approximately considered a zero vector. Due to Step 2), the samples in Class 1 made very little contribution to magnify the eigenvector. Next, the algorithm will iterate over samples in Class 2.

$x_i = (d_{i1}, d_{i2}, \ldots, d_{in})$, $i = s + 1, s + 2, \ldots, s + t$.

Based on Eq. (3.1), y_i is the projection of x_i on the direction of e^{i-1} and $e^{i-1} = e^s$ at this particular moment. We will have:

$$e^{s+t} = \sum\nolimits_i \phi(y_{s+t} x_{s+t}) = y_{s+t} x_{s+t} + \ldots + y_{s+2} x_{s+2} + y_{s+1} x_{s+1} + e_s \tag{3.9}$$

Since e^s is tiny,

$$e^{s+t} \approx \sum\nolimits_i y_i x_i,\ i = s + 1, s + 2, \ldots, s + t \tag{3.10}$$

Depending on the initial angle between e^s and x_{s+1}, e^i and x^i will have the same or opposite direction and the magnitude of e^i is always increasing before normalization.

$$\left\| e^i \right\| > \left\| e^{i-1} \right\|, \ i = s+1, \ s+2, \ldots, \ s+t \tag{3.11}$$

More importantly, the trajectory evolution of e^i will be moving towards Class 2 in this n-dimensional space. The more aggregated the Class 2 is, the more linear the trajectory evolution will be.

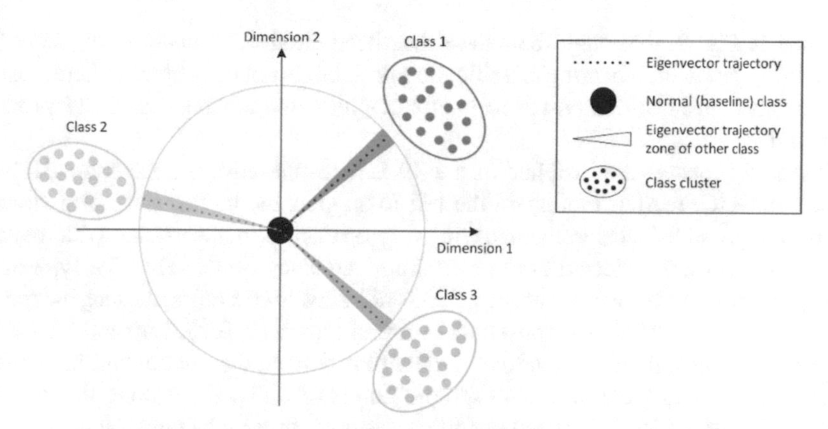

Fig. 1. 2-dimensional graphical illustration of the proposed algorithm

A 2-dimensional graphical illustration of the proposed algorithm is depicted in Fig. 1. Due to Step 2) in Sect. 3.1, data samples in normal (baseline) class are located inside the black circle. As data from a second class come in, the eigenvector e^i gets updated and moves away from O and towards the data sample used to update itself. If the data samples in the second class are quite aggregated, the eigenvector trajectory can be approximated to a line pointing to the center of the second class. Taking Class 2 (green color) in Fig. 1 for example, the eigenvector is updated and formed approximately into a linear eigenvector trajectory towards Class 2's cluster. We use a sector to represent eigenvector trajectory zone that gives some tolerance for its eigenvector trajectory of any future data. The tolerance can be pre-defined and quantified by an angle which can be calculated by:

$$\theta = \cos^{-1}\left(\frac{e_1 \cdot e_2}{\|e_1\| \cdot \|e_2\|} \right) \tag{3.12}$$

4 Results

In this section, the proposed algorithm is applied to a multi-fault bearing classification problem to demonstrate its effectiveness. The vibration data set used in this work is acquired from a standard rolling element bearing test [17]. A diagram of the test rig

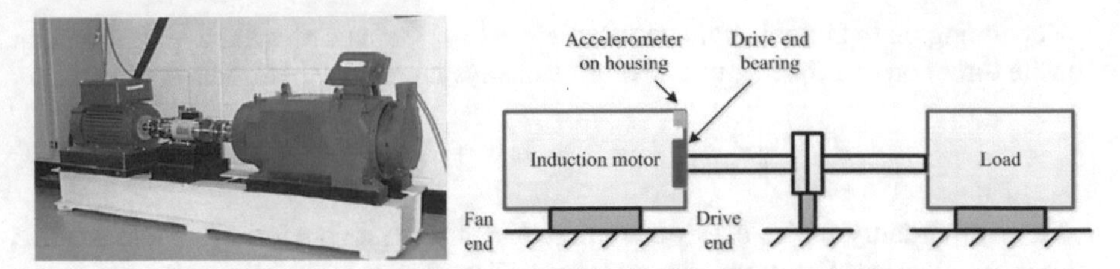

Fig. 2. Test rig for rolling element bearings data acquisition

is shown in Fig. 2. Although the dataset has been studied by many researchers [18–20], solving multi-fault bearing classification problem is not our objective here. Instead, we use this dataset to demonstrate the effectiveness and advantages of our proposed algorithm.

The ball bearings are installed in a 1.49 KW, 3-phase induction motor (Reliance Electric 2HP IQPreAlert motor) on the left to support the motor shaft. The vibration data was acquired by accelerometers, which were attached to the housing with magnetic bases. The data are collected at the sampling frequency of 12 kHz. The type of test bearings is the deep groove ball bearing 6205-2RS JEM SKF. During the test, the rotation speed is at about 30 Hz (1800 rpm) under different loads (0, 0.75, 1.49 and 2.24 KW).

In our analysis, there are totally 472 data blocks from the normal and fault-seeded bearings. Each data block has 2048 sample points (see Fig. 3). Among the data sets, there are 118 data blocks for each condition (normal, inner race fault, outer race fault, and ball fault). Faults are introduced into either the inner raceways or the outer raceways of the drive-end bearings by the electric discharge machining method. The diameters of the faults are 0.1778 mm with depth of 0.2794 mm.

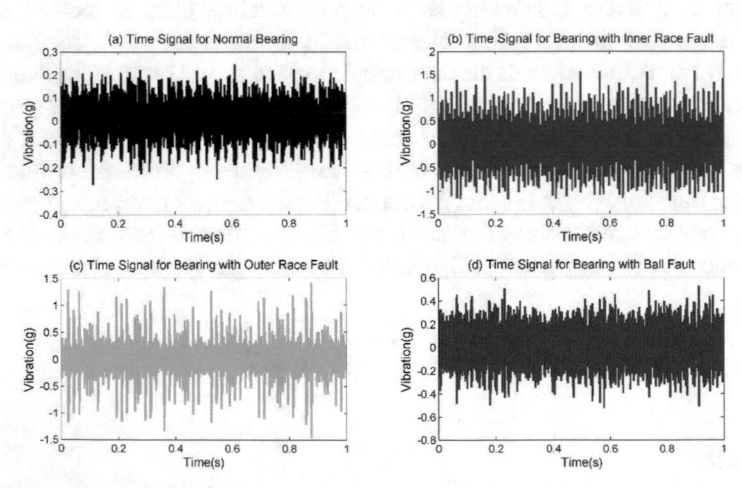

Fig. 3. Time domain vibration signals: (a) raw signals from normal bearing; (b) raw signals from bearing with inner race fault; (c) raw signals from bearing with outer race fault; (d) raw signals from bearing with ball fault.

First, we apply a quick and effective feature extraction method using Wavelet Packet Decomposition (WPD) [21]. 3 level wavelet decomposition is selected so that the feature matrix is 472-by-8. We will not go over the technical details of WPD since it is not the focus of this paper.

Second, we define the following online scenario:

(1) Initially, motor bearing runs in its normal operating condition as our baseline Class 1.
(2) One type of faulty data come in as Class 2. Here we skip the degradation process for simplicity.
(3) As the proposed model processes data continuously, we monitor the eigenvector trajectory of the feature matrix.

Step (1)–(3) are repeated for all three types of faults and Fig. 4 shows their 1st eigenvector trajectories in 8 dimensions.

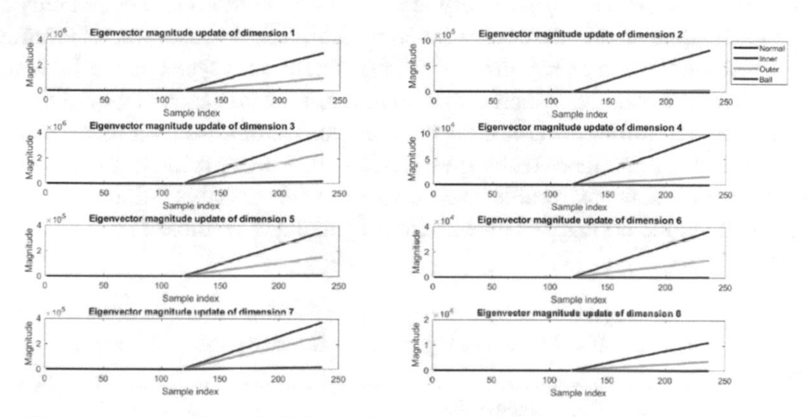

Fig. 4. Eigenvector trajectories in 8 dimensions for four bearing conditions (normal, inner race fault, outer race fault, and ball fault)

A separate Fig. 5 is shown for the zoomed difference of the eigenvector trajectories between normal condition in black and ball fault in red. As shown in Figs. 4 and 5, we have several observations:

(a) The eigenvector has 8 dimensions so we can see their trajectory evolution in 8 subplots.
(b) As expected, data in normal condition has little effect on eigenvector in each dimension.
(c) The eigenvector trajectories of all three fault types can be considered approximately linear.
(d) Before normalization, the magnitude of each eigenvector trajectory increases monotonically.
(e) For different fault types, their eigenvector trajectories have different slopes in each dimension.

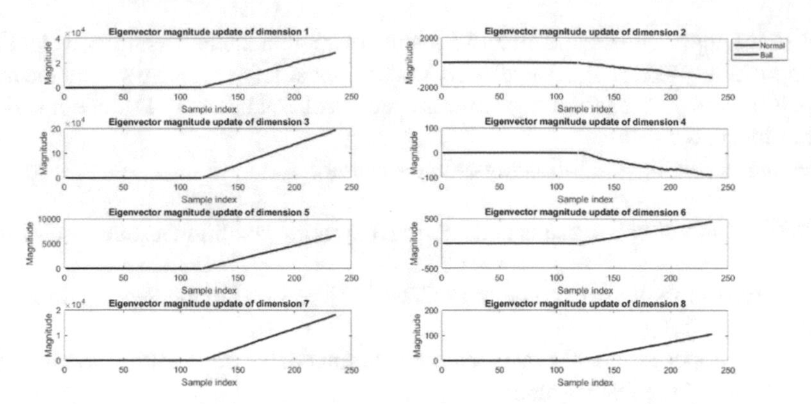

Fig. 5. Zoomed eigenvector trajectories in 8 dimensions for two bearing conditions (normal and ball fault)

In particular, (c), (d), and (e) are very important observations as they not only correspond to the mathematical inferences proved in Sect. 3.2, but also we can leverage the slopes in each dimension as key indicators of different fault types. It can be imagined that if eigenvector's magnitude increases with different rates in each of the 8 dimensions, the eigenvector will evolve towards different directions in this 8-dimensional space. This is because the data of all three fault types locate in different positions relative to O and each eigenvector will move towards its corresponding class in this 8-dimensional space. As an alternative, the normalized eigenvectors (calculated in Table 1) can also be used as key indicators.

Table 1. Normalized reference eigenvectors for each class except normal

Class	Reference eigenvector
Inner race fault	[0.5883, 0.1628, 0.7853, 0.0196, 0.0695, 0.0072, 0.0739, 0.0022]
Outer race fault	[0.3891, 0.0121, 0.9137, 0.0064, 0.0593, 0.0053, 0.0999, 0.0015]
Ball fault	[0.1435, –0.0059, 0.9847, –0.0004, 0.0338, 0.0023, 0.0929, 0.0005]

Next, we proceed to validate the reference eigenvectors in Table 1 can be used as key class indicators. We use new testing data from each class and implement the proposed algorithm in the defined online scenario once again so that Eq. (3.12) can be applied to calculate the angle between reference eigenvectors and new eigenvectors. Results can be seen in Table 2 that there exists an angle between the new eigenvector and reference eigenvector for each class. Among those, ball fault class has the smallest angle and inner race fault class has the largest, which indicate that the data pattern variation of ball fault class is smaller than that of inner race fault class. Also, all angles are below 1.5° so we can set the tolerance discussed in Fig. 1 according to this. Therefore, it has been validated that eigenvector trajectory can be used as key indicators for multi-class classification.

Table 2. Normalized new eigenvectors for each class except normal

Class	New eigenvector	Angle (degree) against reference
Inner race fault	[0.5998, 0.1512, 0.7795, 0.0192, 0.0653, 0.0080, 0.0707, 0.0020]	1.0289
Outer race fault	[0.3796, 0.0119, 0.9183, 0.0059, 0.0544, 0.0053, 0.0972, 0.0013]	0.6887
Ball fault	[0.1429, −0.0059, 0.9848, −0.0005, 0.0338, 0.0023, 0.0930, 0.0005]	0.0215

Further, the proposed algorithm is compared with other legacy machine learning algorithms in Table 3. We run our programs in MATLAB R2017b on a DELL PC with an Intel Core (TM) i7 processer, 4 cores, 1.80 GHz, and 16 GB RAM. It is noticeable that the proposed method requires 22% or even less data processing time and obtains the same classification accuracy on new testing data. Moreover, the proposed algorithm does not require explicit training before online processing as it can adapt to new data pattern when abnormal data comes in. This adaptive property allows itself to perform self-detection and self-classification in online data processing scenario. Depending on whether multi-class data is available, it also gives us tremendous flexibility to selectively perform self-detection or self-classification using the same algorithm, whereas two separate legacy machine learning algorithms need to be trained for anomaly detection and classification respectively.

Table 3. Algorithm benchmark

Algorithm	Averaged processing time for various number of runs (10, 20, 50, 100)	Classification accuracy on new testing data	Require explicit training before online processing?	Adaptive to new patterns?
Decision tree	0.9371 s	1.0	Yes	No
SVM	0.6649 s	1.0	Yes	No
Proposed algorithm	0.1594 s	1.0	No	Yes

5 Conclusion

In this paper, we propose an adaptive self-detection and self-classification algorithm using matrix eigenvector trajectory. This algorithm's mathematical inferences are also introduced and proved theoretically. To demonstrate its effectiveness, this algorithm is applied to a multi-class bearing faults classification problem. In the defined online data processing scenario, self-detection can be realized by monitoring if eigenvector trajectory deviates from the direction of baseline class. Meanwhile, we validate that eigenvector trajectory can be used as key indicators for self multi-class classification.

This algorithm is light-weight, efficient, and accurate. Due to its adaptability, this algorithm does not require explicit training with large amount of data, and it can achieve self-detection and self-classification in an online data processing scenario. Also, this algorithm is not domain specific. Besides machine condition monitoring, it is applicable to other applications as well. For instance, self-detection and self-classification for fake news, spam email, or fraud, etc.

In future work, we will explore more use cases and deploy this algorithm to terminal devices to help people more practically.

References

1. Wijethilaka, S., Liyanage, M.: Survey on network slicing for Internet of Things realization in 5G networks. IEEE Commun. Surv. Tutor. **23**(2), 957–994 (2021)
2. Khuntia, M., Singh, D., Sahoo, S.: Impact of Internet of Things (IoT) on 5G. In: Mishra, D., Buyya, R., Mohapatra, P., Patnaik, S. (eds.) Intelligent and Cloud Computing, vol. 153, pp. 125–136. Springer, Singapore (2021). https://doi.org/10.1007/978-981-15-6202-0_14
3. http://blog.profmobile.com/what-is-internet-of-things
4. Sarker, I.H.: Machine learning: algorithms, real-world applications and research directions. SN Comput. Sci. **2**(3), 1–21 (2021). https://doi.org/10.1007/s42979-021-00592-x
5. Artificial intelligence and machine learning: the next generation (2019)
6. https://brighterion.com/next-generation-artificial-intelligence-machine-learning
7. Basha, A.J., Balaji, B.S., Poornima, S., Prathilothamai, M., Venkatachalam, K.: Support vector machine and simple recurrent network based automatic sleep stage classification of fuzzy kernel. J. Ambient Intell. Humaniz. Comput. **12**(6), 6189–6197 (2020). https://doi.org/10.1007/s12652-020-02188-4
8. Gao, C., Elzarka, H.: The use of decision tree based predictive models for improving the culvert inspection process. Adv. Eng. Inform. **47**, 101203 (2021)
9. Tervonen, J., et al.: Personalized mental stress detection with self-organizing map: from laboratory to the field. Comput. Biol. Med. **124**, 103935 (2020)
10. Wang, Y., Guo, Y.: Forecasting method of stock market volatility in time series data based on mixed model of ARIMA and XGBoost. China Commun. **17**(3), 205–221 (2020)
11. Entezami, A., Sarmadi, H., Behkamal, B., Mariani, S.: Big data analytics and structural health monitoring: a statistical pattern recognition-based approach. Sensors **20**(8), 2328 (2020)
12. Mahmoudi, M.R., Baleanu, D., Mansor, Z., Tuan, B.A., Pho, K.-H.: Fuzzy clustering method to compare the spread rate of Covid-19 in the high risks countries. Chaos Solitons Fractals **140**, 110230 (2020)
13. Yu, C., Han, R., Song, M., Liu, C., Chang, C.-I.: A simplified 2D-3D CNN architecture for hyperspectral image classification based on spatial–spectral fusion. IEEE J. Sel. Top. Appl. Earth Obs. Remote Sens. **13**, 2485–2501 (2020)
14. Chen, H., et al.: A deep learning CNN architecture applied in smart near-infrared analysis of water pollution for agricultural irrigation resources. Agric. Water Manag. **240**, 106303 (2020)
15. Wang, Q., Bu, S., He, Z.: Achieving predictive and proactive maintenance for high-speed railway power equipment with LSTM-RNN. IEEE Trans. Industr. Inf. **16**(10), 6509–6517 (2020)

16. Partridge, M., Calvo, R.A.: Fast dimensionality reduction and simple PCA. Intell. Data Anal. **2**(3), 203–214 (1998)
17. http://www.eecs.case.edu/laboratory/bearing
18. Gryllias, K.C., Antoniadis, I.A.: A support vector machine approach based on physical model training for rolling element bearing fault detection in industrial environments. Eng. Appl. Artif. Intell. **25**(2), 326–344 (2012)
19. Huang, Y., Liu, C., Zha, X.F., Li, Y.: A lean model for performance assessment of machinery using second generation wavelet packet transform and Fisher criterion. Expert Syst. Appl. **37**(5), 3815–3822 (2010)
20. Ocak, H., Loparo, K.A.: Estimation of the running speed and bearing defect frequencies of an induction motor from vibration data. Mech. Syst. Sig. Process. **18**(3), 515–533 (2004)
21. Eren, L., Devaney, M.J.: Bearing damage detection via wavelet packet decomposition of the stator current. IEEE Trans. Instrum. Meas. **53**(2), 431–436 (2004)

Super-Resolution Virtual Scene of Flight Simulation Based on Convolutional Neural Networks

Jiahao Li[1] and Wai Kin Chan[2(✉)]

[1] Interactive Media Design and Technology Center, Tsinghua University, Beijing, China
lijiahao21@mails.tsinghua.edu.cn

[2] Intelligent Transportation and Logistics Systems Laboratory, Tsinghua University, Beijing, China
chanw@sz.tsinghua.edu.cn

Abstract. We propose a method to establish super-resolution (SR) virtual scenes based on multi-spectral remote sensing images. Multi-spectral remote sensing images are processed, then the SR scenes are realized based on the convolutional neural network (CNN) with a special training set. The results show that the training set proposed in this paper can improve the generalization ability of CNN in processing remote sensing images, and different operation sequences can significantly affect the restoration quality of images. The terrain model is established in the physics engine based on the elevation data. Moreover, the flight simulation software with real-time SR virtual scenes is designed and implemented.

Keywords: Super-resolution · Multi-spectral · Virtual scene · CNN

1 Introduction

Flight simulation has been widely used in the multi-media domain, such as pilot training, game interaction, virtual reality scenes, aircraft design, and combat training, etc. [1, 2]. The authenticity and real-time performance of virtual visuals are two aspects that need to be paid attention to in flight simulation [3].

At present, the research in this field mainly focuses on the real-time modeling of 3D terrain [4], to solve the problem of making textures and 3D models have high realism. There is few research on the processing of real remote sensing images and their application in virtual vision. For the real-time visualization of 3D models, the current research mainly explores the improvement of hardware and the efficiency of graphics rendering algorithms [1, 4, 5, 20]. LOD technology, data storage, and compression technology are commonly used [5, 6], but the artificial neural networks are rarely used for flight simulation.

The reflection of different materials on the ground is very complex, so it is difficult to accurately describe all the light reflection processes by real-time rendering [5, 6]. Therefore, there is still a large gap between the virtual scenes using optical 3D models and human visual effects. Remote sensing images are taken by sensors, which already

© The Author(s), under exclusive license to Springer Nature Switzerland AG 2023
L. C. Tang and H. Wang (Eds.): BDET 2022, LNDECT 150, pp. 138–147, 2023.
https://doi.org/10.1007/978-3-031-17548-0_13

contain a lot of ground reflection information. Firstly, this paper explores the processing method and technical route of remote sensing images applied to virtual scenes, then carries out SR restoration of virtual scenes based on CNN, and puts forward a loading scheme of real-time visualization scenes. This paper designs and implements a flight simulation system in the PC platform. The experimental results show that the method proposed in this paper has high practical application value.

2 Multi-spectral Remote Sensing Image Processing

To establish the real virtual scenes outside the aircraft window, this paper selects the multi-spectral remote sensing images taken by earth resources satellites as the materials of the virtual visual scenes. The lighting model of virtual modeling is an approximate simplification of natural lighting, which is difficult to simulate the real reflection of ground objects. However, remote sensing images can truly reflect the characteristics of ground objects. Remote sensing images have met the real lighting conditions, which do not need complex lighting rendering calculation and have a higher sense of reality.

2.1 Multi-band Remote Sensing Image Synthesis

This paper mainly selects Landsat series satellites remote sensing images for the experiment. The wide range of high-definition image data of Landsat series satellites provides a training set for the convolutional neural network and provides experimental conditions of the super-resolution scenes. Landsat series satellites are equipped with MSS, TM, ETM+, OLI, and other types of sensors.

Remote sensing images color synthesis is to give different colors to remote sensing images in different bands, that is, different bands are combined and given three primary colors to synthesize visually colored images.

According to the Helmholtz theory of human vision [7], 3-2-1 band is selected as the spectral band of band synthesis. The real color effects synthesized in this paper are shown in Fig. 1.

Fig. 1. 3-2-1 band synthesis

2.2 Atmospheric Correction

Due to the influence of the earth's atmosphere on electromagnetic radiation in the process of remote sensing images acquisition, the propagation of electromagnetic radiation between ground objects and sensors is affected, and the transmission speed, intensity, wave length, and spectrum of electromagnetic radiation are changed. Therefore, remote sensing image data need to be preprocessed, including remote sensing detector system error correction, atmospheric correction, slope aspect influence correction, geometric correction.

Atmospheric correction algorithms mainly include ACORN, ATREM, FLAASH, ATCOR, etc.

ACORN is based on MODTRAN (Alder-Golden, 1999) radiation transfer model [8], and the formula is as follows:

$$L_s = \frac{E_{o_\lambda}\left[E_{du_\lambda} + \frac{T_{\theta_o}\rho_\lambda T_{\theta_v}}{1-E_{dd_\lambda}\rho_\lambda}\right]}{\pi} \quad (1)$$

where L_s is the total radiation received by the detector; E_{o_λ} is the spectral irradiance of the sun; E_{du_λ} and E_{dd_λ} are atmospheric reflections in the upper and lower directions respectively; T_{θ_o} and T_{θ_v} are the atmospheric transmission in the upper and lower directions respectively; ρ_λ is the converted surface spectral reflectance.

Therefore, the converted surface spectral reflectance is:

$$\rho_\lambda = \frac{1}{\left[\frac{(E_{o_\lambda}T_{\theta_o}T_{\theta_v})/\pi}{L_S-(E_{o_\lambda}E_{du_\lambda})/\pi}\right] + E_{dd_\lambda}} \quad (2)$$

In this paper, ENVI software is used for atmospheric correction. Remote sensing images processing effectively reduces or eliminates the errors in the process of images acquisition, which provides a basis for further flight simulation application.

3 Super-Resolution Restoration

There are two problems in the direct application of remote sensing images to virtual scenes: 1) The normal resolution of remote sensing images can not meet the requirements of virtual vision. Take the Landsat as an example, the resolution of the Landsat MSS sensor is 80 m, that of Landsat TM and ETM + sensor is 30 m (Band 6 is 60 m, ETM + band 8 is 15 m), and that of Landsat OLI sensor is 30 m. Remote images are difficult to meet the human visual requirements of flight simulation. 2) It is difficult for personal computers to store and load a large amount of remote sensing images data of virtual scenes. In this paper, the SRCNN is used to construct the super-resolution virtual scenes of flight simulation.

3.1 Super-Resolution and SRCNN

Super-resolution restoration is a technology that uses computer algorithms to enhance image resolution, that is to restore low resolution (LR) images and establish super-resolution (SR) images [9, 10]. The super-resolution of remote sensing images is not the simple enlargement of the images, but the restoration or reconstruction of the detailed information in the images.

Super-resolution restoration is usually carried out from the following perspectives: 1) Reconstruction-based super-resolution restoration; 2) Learning-based super-resolution restoration; 3) super-resolution restoration based on other methods. There are two kinds of image super-resolution restoration methods based on reconstruction: frequency domain and spatial domain. The representative algorithms of spatial domain based super-resolution restoration methods include non-uniform interpolation (NUI) algorithm, MAP algorithm, POCS algorithm. Learning-based image super-resolution technology mainly includes shallow learning and deep learning. Super-resolution restoration methods based on deep learning mainly include generative adversarial network (GAN), sparse coding based network (SCN), convolution neural network (CNN), and deep belief network (DBN).

Table 1. Average PSNR of SRCNN

CNN \ Magnification / PSNR	3	4	6
9-1-5(91 images)	19.93	18.72	17.58
9-1-5(ImageNet)	19.98	18.73	17.59
9-5-5(ImageNet)	20.02	18.73	17.59

After comparing various methods, this paper uses the convolutional neural networks which are most suitable for flight simulation to construct the super-resolution virtual scene. Dong C et al. proposed the application of CNN to image super-resolution restoration [11]. We choose 9-1-5 and 9-5-5 to do experiments on SRCNN on different training sets. 5 Landsat remote sensing images are selected for testing, and the average results are shown in Table 1.

The results show that the generalization ability of CNN trained in Imagenet to remote sensing images is very poor. PSNR shows that these images cannot be used in virtual scenes. It can be seen from the results that with the increase of magnification, there is almost no gap between 9-1-5 and 9-5-5 (accurate to two decimal places), which is consistent with the conclusion in Dong's paper [11], so the subsequent experiments in this paper are only carried out on 9-1-5. We changed the training sets and carried out the experiment. The ImageNet and 91 images are tested [11]. The results show that different training sets such as ImageNet have little impact on PSNR, which means that because these training sets do not consider special images such as remote sensing images, they can not be directly used in super-resolution restoration of special images.

This paper focuses on solving two key problems in the application of convolutional neural networks in virtual scenes: 1) Remote sensing images are different from usual images. The effect of convolutional neural networks directly applied to remote sensing images super-resolution restoration is poor. In this paper, high-definition remote sensing images are introduced into the training sets of neural network to make the restoration results more accurate. 2) The neural networks trained in this paper is not just to verify the effect through fixed data sets, but to be applied to multimedia software. Therefore, the generalization ability of neural networks is of vital importance.

3.2 Training Set

The establishment of the training set must be able to provide sufficient prior knowledge in the field of remote sensing images for convolutional neural network [12, 13]. The data sets in this paper does not use the common down-sampling method [14, 15], but uses the method based on different scales. As shown in Fig. 2, first sample the remote sensing images with a scale of 1:5, and then sample the images with a scale of 1:30, so that the distribution of pixels obtained is closer to human vision. Due to the great differences of ground objects' reflection in different seasons, this paper selects the remote sensing images of each season to improve the accuracy. In this paper, a total of 120 pairs of low-resolution and high-resolution images are produced as the training set (Automatically divided by MATLAB program according to coordinates). The high-resolution images are 270 × 270, and the low-resolution images are 45 × 45.

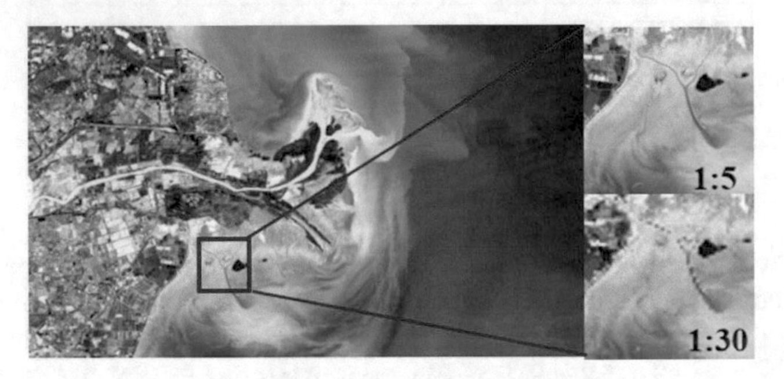

Fig. 2. Process of training set making

3.3 Super-resolution Remote Sensing Images

Image is a kind of data set with three dimensions, so the operation of data shall be based on three channels [16]. To make the training effect of convolutional neural networks better, this paper first converts RGB images into YCbCr images, as shown in Fig. 3.

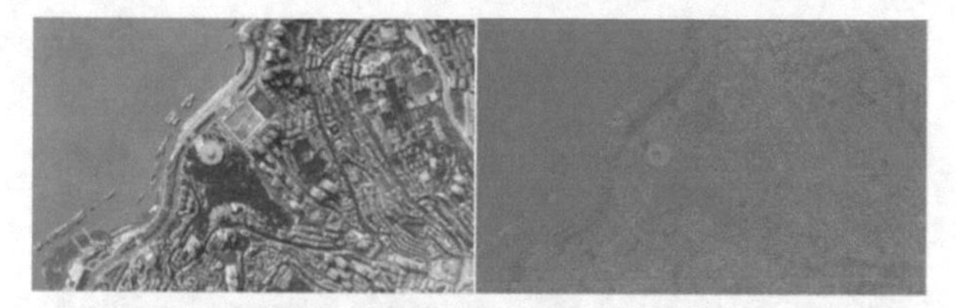

Fig. 3. RGB and YCbCr

Table 2. Average PSNR on the special training set

Magnification / PSNR / CNN	3	4	6
9-1-5	26.65	25.04	23.61

The PSNR results of 9-1-5 SRCNN on the training set proposed in this paper are shown in Table 2. The super-resolution restoration of remote sensing images is shown in Fig. 4. The results combining PSNR and visual effect show that the images obtained in this paper can be applied to the super-resolution scenes of flight simulation.

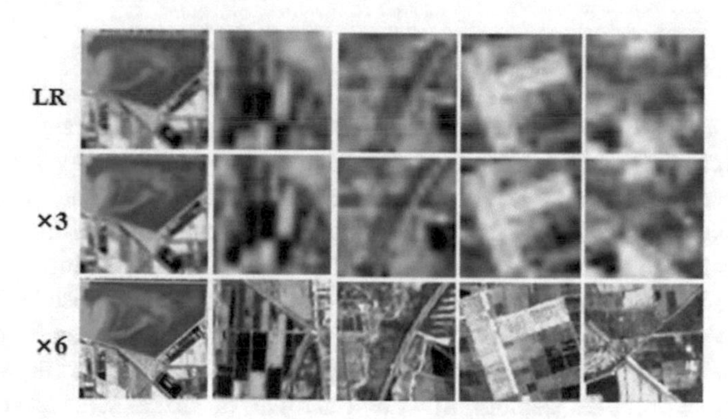

Fig. 4. Super-resolution restoration

4 Real-Time Visualization and Virtual Scene

In this paper, a real-time visualization method based on physics engine [17, 18, 19] and SRCNN is proposed, and the flight simulation software is implemented.

4.1 Terrain Model Based on Elevation Data

Physics engine has been widely used in games, multimedia, engineering, and other fields [20, 21]. Unity is a widely used physics engine. Firstly, this paper carries out

Fig. 5. Terrain elevation data and 3D models

terrain modeling based on elevation data through Unity engine. The terrain models are established by the grid points composed of the elevation data. The terrain data are obtained through the elevation satellites, and then the elevation data is imported into the Unity physics engine for terrain modeling, as shown in Fig. 5.

4.2 Real-Time Loading of Super-Resolution Scene

In this paper, a real-time visualization method based on physics engine and convolutional neural networks is proposed. The box colliders are set up in Unity physics engine, and the collision detection module is implemented by C# program. The position of the aircraft is obtained in real-time through the collision trigger events at different positions. The position of the aircraft is closely related to the real-time loading of the scenes. According to different aircraft positions, super-resolution scenes are used within certain fields of view of the aircraft, and ordinary interpolation scenes are used for other fields. The real-time loading scheme can reduce the difficulty of scenes loading and improve the efficiency of software without affecting the effect of virtual scenes. The real-time visualization process is shown in Fig. 6.

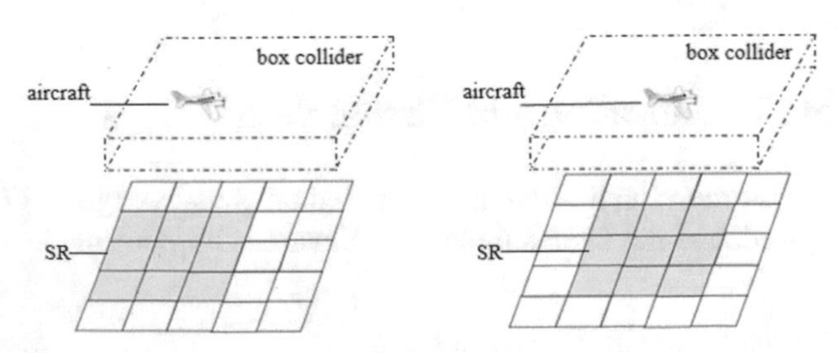

Fig. 6. Real-time loading process

4.3 Virtual Scene Implementation

We designed and implemented a flight simulation software, and evaluated the visual effect. Although the low-resolution and super-resolution images look similar when the image size is small, there is a large gap in visual effect when the images are enlarged,

Fig. 7. Comparison between SR and ordinary scenes

especially when applied to virtual 3D scenes. As shown in Fig. 7, the super-resolution scenes significantly improve the visual effect. The details in the scenes are well restored, and the resolution of mountain textures, river tributaries and vegetation details are significantly improved. Even if these super-resolution images processed through convolutional neural networks are applied to the establishment of 3D virtual scenes, good results are achieved, which shows that the convolutional neural network based on special training set has good generalization ability.

Fig. 8. Atmospheric reflection and fog rendering

5 Conclusion and Future Work

In this paper, we proposed an implementation method of super-resolution virtual scenes, which improved the generalization ability of SRCNN in the application of virtual scenes based on remote sensing images. Based on the real elevation data, we carried out the 3D terrain modeling in the physics engine, established the virtual scenes using super-resolution remote sensing images, and designed a flight simulation software. We evaluated the results from two aspects: PSNR and human vision. The evaluation shows that the method proposed in this paper has achieved good results.

Figure 8 shows the high-altitude fog rendering effect in the physics engine. The physics engine can render virtual scenes and adjust the parameters according to different conditions. Therefore, it can be further explored to use the physics engine to further render virtual scenes and adjust the scenes according to the characteristics of different aircraft and weather.

Acknowledgements. This research was funded by the National Natural Science Foundation of China (Grant No. 71971127), Guangdong Pearl River Plan (2019QN01X890), and Shenzhen Science and Technology Innovation Commission (JCYJ20210324135011030).

References

1. Ortiz, G.A.: Effectiveness of PC-based flight simulation. Int. J. Aviat. Psychol. **4**(3), 285–291 (1994)

2. Baarspul, M.: A review of flight simulation techniques. Prog. Aerosp. Sci. **27**(1), 1–120 (1990)
3. Wang, W., Li, D., Liu, C.: Helicopter flight simulation trim in the coordinated turn with the hybrid genetic algorithm. Proc Inst. Mech. Eng. Part G: J. Aerosp. Eng. **233**(3), 1159–1168 (2019)
4. Huang, M.Q., et al.: Multi-LOD BIM for underground metro station: interoperability and design-to-design enhancement. Tunn. Undergr. Space Technol. **119**, 104232 (2022)
5. Xiong, C., Lu, X., Huang, J., Guan, H.: Multi-LOD seismic-damage simulation of urban buildings and case study in Beijing CBD. Bull. Earthq. Eng. **17**(4), 2037–2057 (2018). https://doi.org/10.1007/s10518-018-00522-y
6. Wu, J., et al.: Research on high precision terrain dynamic loading technology based on flight trajectory prediction. In: 2018 IEEE 4th Information Technology and Mechatronics Engineering Conference (ITOEC). IEEE (20180)
7. Dinguirard, M., Slater, P.N.: Calibration of space-multispectral imaging sensors: a review. Remote Sens. Environ. **68**(3), 194–205 (1999)
8. Sun, W., et al.: A band divide-and-conquer multispectral and hyperspectral image fusion method. IEEE Trans. Geosci. Remote Sens. **60**, 1–13 (2021)
9. Lv, Y., Ma, H.: Improved SRCNN for super-resolution reconstruction of retinal images. In: 2021 6th International Conference on Intelligent Computing and Signal Processing (ICSP). IEEE (2021)
10. Sujith Kumar, V.: Perceptual image super-resolution using deep learning and super-resolution convolution neural networks (SRCNN). Intell. Syst. Comput. Technol. **37**(3) (2020)
11. Dong, C., et al.: Learning a deep convolutional network for image super-resolution. In: Fleet, D., Pajdla, T., Schiele, B., Tuytelaars, T. (eds.) Computer Vision, vol. 8692, pp. 184–199. Springer, Cham (2014). https://doi.org/10.1007/978-3-319-10593-2_13
12. Park, S.C., Park, M.K., Kang, M.G.: Super-resolution image reconstruction: a technical overview. IEEE Sig. Process. Mag. **20**(3), 21–36 (2003)
13. Chaudhuri, S. (ed.): Super-Resolution Imaging, vol. 632. Springer, Cham (2001). https://doi.org/10.1007/b117840
14. Dong, C., et al.: Image super-resolution using deep convolutional networks. IEEE Trans. Pattern anal. Mach. Intell. **38**(2), 295–307 (2015)
15. Da Wang, Y., Armstrong, R.T., Mostaghimi, P.: Enhancing resolution of digital rock images with super resolution convolutional neural networks. J. Pet. Sci. Eng. **182**, 106261 (2019)
16. Ooyama, S., et al.: Underwater image super-resolution using SRCNN. In: International Symposium on Artificial Intelligence and Robotics 2021, vol. 11884. SPIE (2021)
17. Liu, X.: Three-dimensional visualized urban landscape planning and design based on virtual reality technology. IEEE Access **8**, 149510–149521 (2020)
18. Sun, S., et al.: Application of virtual reality technology in landscape design. In: 2021 International Symposium on Artificial Intelligence and its Application on Media (ISAIAM). IEEE (2021)
19. Lu, G.P., Xue, G.H., Chen, Z.: Design and implementation of virtual interactive scene based on unity 3D. In: Advanced Materials Research, vol. 317. Trans Tech Publications Ltd (2011)
20. Kuang, Y., Bai, X.: The research of virtual reality scene modeling based on unity 3D. In: 2018 13th International Conference on Computer Science and Education (ICCSE). IEEE (2018)

User Assignment for Full-Duplex Multi-user System with Distributed APs

Ying Shen[1], Juan Zhou[2(✉)], and Wenzao Li[2]

[1] National Key Laboratory of Science and Technology on Communications, University of Electronic Science and Technology of China, No. 2006, Xiyuan Avenue, West Hi-Tech Zone, Chengdu, China

[2] College of Communication Engineering, Chengdu University of Information Technology, No. 24 Block 1, Xuefu Road, Chengdu, China

zhoujuan@uestec.edu.cn

Abstract. Full duplex (FD) communication device transmits and receives wireless signals at the same time and at the same frequency, which could double the spectral efficiency. This paper focuses on the user assignment for a multi-user and multi-antenna cellular network where the FD base station (BS) simultaneously communicates with half-duplex (HD) uplink and downlink users. From the generalized degree of freedom (GDoF), we provide an optimal user assignment method for the FD multi-user multi-antenna network. The optimality problem is formulated as a linear program for bipartite graph. Furthermore, we prove that the optimal user assignment can be found by solving two maximum weighted matching problems in polynomial time. Simulation results show that the proposed user assignment achieves significant sum throughput improvement.

Keywords: User assignment · Dull duplex · Distributed APs

1 Introduction

Full duplex (FD) communication has attracted lots of attention because of its potential to double the spectrum efficiency, by achieving transmission and reception on the same carrier frequency, simultaneously [1–3]. In this paper, we consider a multi-user and multi-antenna cellular network. In the considered network, the FD base station (BS) is equipped with multiple distributed antennas, which simultaneously communicates with multiple HD uplink (UL) and downlink (DL) users (UEs). In the traditional HD systems, the resource allocations of UL and DL are independently performed. While in the FD cellular network, both the UL and DL communications exist at the same time. Therefore, the traditional HD UL or DL user assignment method will inevitably lead to performance loss in the FD cellular network, and the new user assignment method is required.

The existing multi-antenna full-duplex user assignment literatures focus on the centralized multi-antenna base station [4,5]. However, the distributed

© The Author(s), under exclusive license to Springer Nature Switzerland AG 2023
L. C. Tang and H. Wang (Eds.): BDET 2022, LNDECT 150, pp. 148–160, 2023.
https://doi.org/10.1007/978-3-031-17548-0_14

antenna system has the advantages of system capacity and reliability, and it also can reduce the complexity of interference elimination by fully exploiting spatial channel resources without additional spectrum resources and transmit power. Therefore, it is necessary to study the optimal assignment method for users to access the full-duplex network with distributed multi-antennas. Besides, the existing user assignment problem is developed based on the traditional capacity formula with high complexity. To reduce the complexity, we will deal with this optimization problem from the perspective of generalized degrees of freedom (GDoF). From a GDoF perspective, we study the optimal user assignment in the FD multi-user multi-antenna network. This optimality problem is formulated as a linear program for bipartite graphs. By solving two maximum weighted matching problems in polynomial time, the optimal user assignment can be obtained. The remainder of this paper is organized as follows. In Sect. 2, we introduce the system model. The proposed optimal user assignment is described in Sect. 3. Section 4 presents numerical results and finally, conclusions are drawn in Sect. 5.

2 System Model

The multi-user multi-antenna FD system is shown in Fig. 1. Here in the FD system under consideration, the FD base station (BS) is equipped with N_a distributed antennas, which can provide multiple access points(APs). The BS simultaneously communicates with N_u HD UL and DL users over one frequency.

One AP communicates with one HD UL user and one HD DL user over one frequency, as shown in Fig. 2. Correspondingly, the FD BS can communicate with $2N_a$ HD UL and DL users over one frequency. The system is supposed to have N_f frequencies. Then, the BS can simultaneously communicate with $2N_aN_f$ HD UL and DL users over N_f frequencies.

Throughout the paper, we assume that over one frequency the UE $N_u \leq 2N_a$; otherwise, there exists at least one UE who can never be assigned to any AP.

2.1 Network Model

All the UEs can be considered as the nodes in this network. Let AP (k, m) denote the k-th AP over the m-th frequency. For the convenient of analysis, the AP over one frequency is also considered as one node, meaning AP (k, m) is one node in the network. Then N_a APs over N_f frequencies represent N_fN_a nodes in the network.

If the number of APs is $N_a = 2$, the number of UEs available on one frequency is $N_u = 4$, and the number of frequencies is $N_f = 2$. Then the total number of UE nodes is $N_fN_u = 8$, and we get the corresponding system network as shown in Fig. 3.

It is worth noting that the total number of nodes equals

$$N_o = N_fN_a + N_fN_u \tag{1}$$

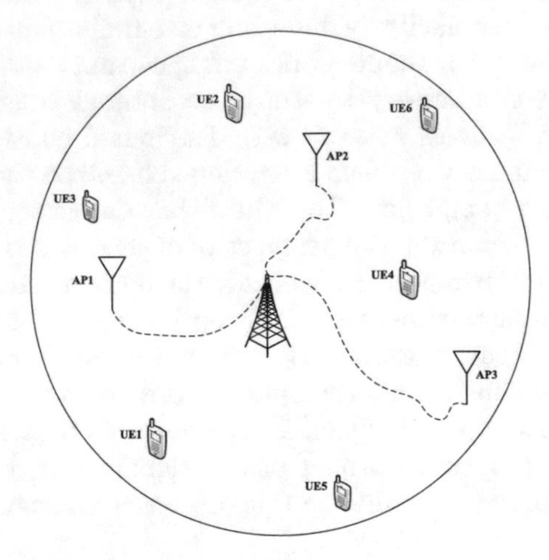

Fig. 1. Multi-user multi-antenna FD system model. AP denotes access point

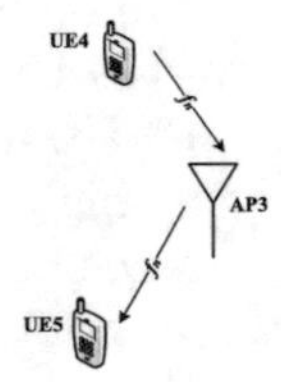

Fig. 2. One AP communicates with two different UEs over one frequency.

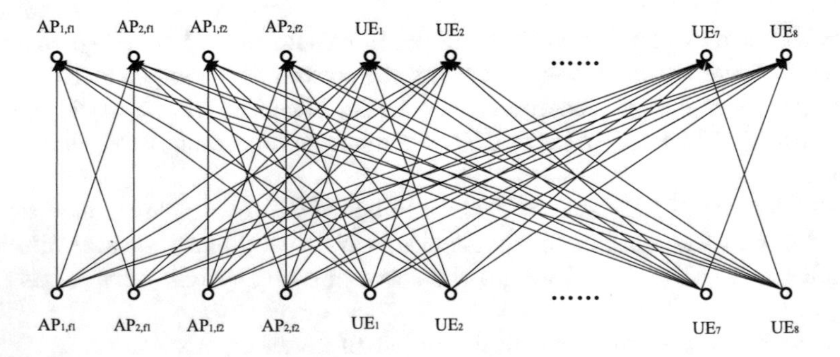

Fig. 3. An example of network with 2 APs and 8 UEs over 2 frequencies.

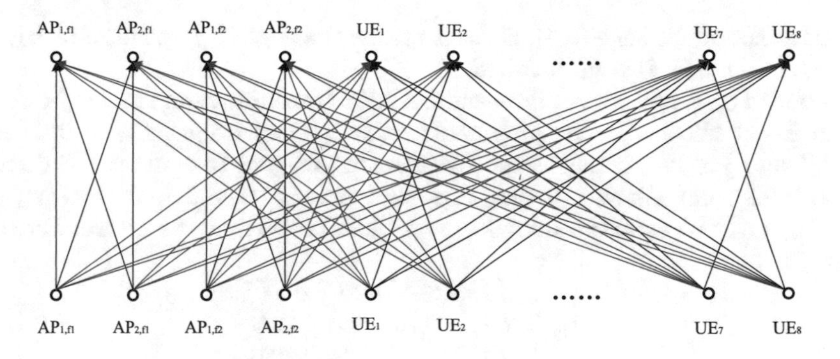

Fig. 4. Links need to be studied.

Due to the symmetry between DL and UL, the target network can be denoted by bipartite graph

$$\mathcal{G} = (\mathcal{K}, \mathcal{K}, \mathcal{E}) \tag{2}$$

with left vertices $\mathcal{K}$, right vertices $\mathcal{K}$, and edges $\mathcal{E} = \mathcal{K} \times \mathcal{K}$, where $\mathcal{K} \triangleq \{1, 2, ..., N_o\}$. A matching $\mathcal{M} \subseteq \mathcal{E}$ is a set of edges. When weights $w(i, j)$ are associated to the edges $(i, j) \in \mathcal{E}$, the weight of the matching $\mathcal{M}$ is defined by

$$w(\mathcal{M}) = \sum\nolimits_{(i,j)\in\mathcal{M}} w(i, j) \tag{3}$$

and the maximum weighted matching, i.e., the matching with maximum sum-weight is then denoted by

$$\mathcal{M}^* = \arg\max\nolimits_{\mathcal{M}} w(\mathcal{M}) \tag{4}$$

For the convenient of analysis, $\mathcal{K}$ can be rewritten as

$$\mathcal{K} \triangleq \{\mathcal{N}, \mathcal{N}^c\} \tag{5}$$

with $\mathcal{N} \triangleq \{1, 2, ..., N_f N_a\}$ and $\mathcal{N}^c \triangleq \{N_f N_a + 1, ..., N_f N_a + N_f N_u\}$.

The relationship between the indexes in $\mathcal{K}$ and nodes is shown as follows:

$$\begin{array}{rccccccc} \mathcal{N} = & \{1, & 2, & \ldots & N_a, & N_a + 1, & \ldots & N_f N_a\} \\ & \updownarrow & \updownarrow & & \updownarrow & \updownarrow & & \updownarrow \\ & AP(1,1) & AP(2,1) & & AP(N_a,1) & AP(1,2) & & AP(N_a,N_f) \end{array} \tag{6}$$

and

$$\begin{array}{rcccc} \mathcal{N}^c = & \{N_a N_f + 1, & N_a N_f + 2, & \ldots & N_a N_f + N_f N_u\} \\ & \updownarrow & \updownarrow & & \updownarrow \\ & UE_1 & UE_2 & & UE_{N_f N_u} \end{array} \tag{7}$$

We say that an edge (i, j) denotes that the node i is allocated to node j over one frequency. In another words, node i transmits signal to node j. For the convenience of analysis, the $N_o \times N_o$ adjacent matrix $\mathbf{B}$ is employed to describe the existence of the edge (i,j). $\mathbf{B}_{ij}=1$ means the edge (i,j) exists, and $\mathbf{B}_{ij}=0$ means the edge (i,j) is not allowed.

If both $i\in\mathcal{N}$ and $j\in\mathcal{N}$, it means the AP i transmits signal to AP j directly, which is not allowed in the proposed network. Then we get $\mathbf{B}_{ij}=0$. If both $i\in\mathcal{N}^c$ and $j\in\mathcal{N}^c$, it means the UE transmits signal to another UE directly, which is also not allowed in the proposed network, then $\mathbf{B}_{ij}=0$. Otherwise, edge (i,j) can be exist, meaning $\mathbf{B}_{ij}=1$. It is noted that the matrix $\mathbf{B}$ can be given by

$$\mathbf{B}_{ij}=\begin{cases}1, & i\in\mathcal{N} \text{ and } j\in\mathcal{N}^c\\ 1, & i\in\mathcal{N}^c \text{ and } j\in\mathcal{N}\\ 0, & \quad\quad others\end{cases} \tag{8}$$

It is noticed that in this paper we focus on the UE-AP pair allocation so as to maximize the GDoF of the network. Hence UE can not be connected with UE directly, and the AP can not be connected with AP directly, either. Then only red lines of the graph in Fig. 4 needs to be studied to achieve the maximum GDoF.

2.2 Channel Model

The system is assumed to has multi-user Gaussian interference channel. At edge (i,j) and $\mathbf{B}_{ij}=1$, the received signal at node j is given by

$$Y_j(t)=\sum_{i\in\mathcal{K}} h_{ij}\widetilde{X_i}(t)+Z_j(t) \tag{9}$$

where $\widetilde{X}_i$ is the transmitted signal from node i with power constraint $E[|\widetilde{X}_i|^2]\le P_i$, h_{ij} is the channel coefficient between node i and node j, and $Z_j(t)\sim\mathcal{CN}(0,1)$ is the (normalized) additive white Gaussian noise at node j.

Following [6,7], the signal model in (9) is translated in to equivalent GDoF-friendly form, given by

$$Y_j(t)=\sum_{i\in\mathcal{K}}\sqrt{P^{\alpha_{ij}}}e^{j\theta_{ij}}X_i(t)+Z_j(t) \tag{10}$$

where $X_i(t)=\frac{\widetilde{X}_i(t)}{\sqrt{P_i}}$ is the normalized transmitted signal from node i with the the corresponding power constraint $\mathbb{E}[|X_i(t)|^2\le 1]$, $\sqrt{P^{\alpha_{ij}}}$ and θ_{ij} are channel gain and phase shifter of the channel coefficient between node i and node j, respectively, and the corresponding channel strength level α_{ij} is defined as

$$\alpha_{ij}=\frac{\log\left(\max\left(1,|h_{ij}|^2P_i\right)\right)}{\log P} \tag{11}$$

where P is the average power for the edge (i, j). The actual transmit power could be denoted as $P^{r_i} P_i$, with a transmit power adjustment defined as P^{r_i}.

The corresponding $N_0 \times N_0$ channel strength level matrix $\mathbf{A}$ can be given by

$$\mathbf{A} = \left[\mathbf{A}_1, \mathbf{A}_2 \ldots \mathbf{A}_{N_a N_f}, \mathbf{A}_{N_a N_f+1} \cdots \mathbf{A}_{N_a N_f+N_u N_f}\right] \tag{12}$$

with

$$\mathbf{A}_i = \left[\alpha_{i1},\ \alpha_{i2} \cdots \alpha_{i(N_a N_f)}, \alpha_{i(N_a N_f+1)} \cdots \alpha_{i(N_a N_f+N_u N_f)}\right]^T \tag{13}$$

$$\mathbf{A}_{ij} = \alpha_{ij} \tag{14}$$

where α_{ij} in $\mathbf{A}$ denotes the channel strength level value of edge (i, j) and $i, j \in \mathcal{K}$.

2.3 Treating Interference as Noise

In a multi-user Gaussian interference channel, the strengths of the strongest interference from an user α is denoted F_α (values in dB scale), and the strengths of the strongest interference to this user is I_α (values in dB scale), the desired signal strength is D_α (values in dB scale). It is shown in [6] that TIN (with power control) is optimal from a GDoF perspective if $D_\alpha \geq I_\alpha + F_\alpha$ can be satisfied for each user, with which the entire capacity area to be within a constant gap can be achieved. Hence, the TIN is applied in the multi-user multi-antenna FD system.

For the convenience of analysis, the achievable rate for message W_{ij} of the edge (i, j) is given as R_{ij}, and the corresponding individual achievable GDoF of message W_{ij} is denoted as $d_{ij} \triangleq \lim_{P\to\infty} \frac{R_{ij}}{\log P}$. Suppose all the users are paired well with the APs. Then the number of the available edge is $K = N_u N_f$.

The GDoF region is the collection of all achievable GDoF-tuples $(d_{i_1 j_1}, d_{i_2 j_2}, \ldots, d_{i_K j_K})$. The TIN-Achievable GDoF (TINA) region defined in [6,7] is the set of all K-tuples $\mathbf{d} = (d_{i_1 j_1}, d_{i_2 j_2}, \ldots, d_{i_K j_K})$, satisfying

$$d_{i_k j_k} \leq \max\left\{0, \alpha_{i_k j_k} + r_{i_k} - \max\left\{0, \max_{n:n\neq k} (\alpha_{i_n j_k} + r_{i_n})\right\}\right\} \tag{15}$$

for some assignment of the power allocation variables $\mathbf{r} = (r_{i_1}, r_{i_2}, \ldots, r_{i_K})$.

From [6,7] the polyhedral TINA region is defined by removing the positive part operator from the right side of (15). It is able to find a convex polytope form for the polyhedral TINA region for any subnetwork formed by a subset $\mathcal{S} \subseteq \mathcal{K}$ and its associated desired and interfering links. Such a polytope is denoted by $\mathcal{P}_{\mathcal{S}}^{\text{TINA}}$.

3 Optimal User Assignment

The optimal assignment scheme to achieve the maximum sum throughput can be formulated as the following maximum achievable GDoF:

$$\max \sum_{i \in \mathcal{K}} \sum_{j \in \mathcal{K}} d_{ij} x\left(i,j\right) \tag{16a}$$

$$\begin{aligned} s.t. & \\ & x\left(i,j\right) \in [0,1]. && (16b)\\ & x\left(i,j\right) \leq \mathbf{B}_{ij},\ i,j \in \mathcal{K} && (16c)\\ & \sum_{i \in \mathcal{K}} x\left(i,j\right) \leq 1,\ j \in \mathcal{N} && (16d)\\ & \sum_{j \in \mathcal{K}} x\left(i,j\right) \leq 1,\ i \in \mathcal{N} && (16e)\\ & \sum_{i \in \mathcal{K}} x\left(i,m\right) + \sum_{j \in \mathcal{K}} x\left(m,j\right) \leq 1,\ m \in \mathcal{N}^c && (16f) \end{aligned}$$

It is known that this linear program is equal to the maximum weight sum over all matchings of the network [8]. Let us model the proposed network using a bipartite graph in $\mathcal{S} \subseteq \mathcal{K}$, the transmitters and receivers are in the two point classes respectively, the desired and interference links are the edges in the graph. By replacing (16b) with $x(i,j) = \{0,1\}$, (16a) can be shown as a maximum weighted matching problem.

Here the constraints (16b) and (16c) mean that the coefficient $x(i,j)$ of d_{ij} can only be 0 or 1. An edge (i,j) belongs to the maximum-weighted matching if $x(i,j) = 1$, i.e., edge (i,j) is available and node i transmits to node j.

The constraint (16d) denotes that node $j (j \in \mathcal{N})$ is an receive AP over one frequency, and node i must be an UE, since only UE and AP can be paired. Furthermore, it means the AP as a receiver over a certain frequency can be only paired with an UE for one time.

Similarly, the constraint (16e) denotes that node $i (i \in \mathcal{N})$ is an transmit AP over one frequency, and node j must be an UE. Besides the AP as a transmitter over a certain frequency can be only paired with an UE for one time.

The constraint (16f) means that if the node m is half-duplex UE over one frequency, then it can not transmit and receive at the same time in the same frequency. Hence the UE over a certain frequency can be only paired with an AP for one time.

Based on the Theorem 3 in [7], for any subset $\mathcal{S} \subseteq \mathcal{K}$, $\mathcal{P}_{\mathcal{S}}^{\text{TINA}}$ is given by

$$\left\{ \begin{array}{l} (d_{ij} : i,j \in \mathcal{K}) : \\ \qquad d_{ij} \geq 0, \forall i,j \in \mathcal{S}, d_{gh} = 0, \forall g,h \in \mathcal{S}^c \\ \sum_{i \in \mathcal{S}'} \sum_{j \in \mathcal{S}'} d_{ij} \leq \sum_{i \in \mathcal{S}'} \sum_{j \in \mathcal{S}'} \alpha_{ij} - w\left(\mathcal{M}_{\mathcal{S}'}^*\right), \forall \mathcal{S}' \subseteq \mathcal{S} \end{array} \right\} \tag{17}$$

where $\mathcal{S}^c$=$\mathcal{K} \backslash \mathcal{S}$. $\mathcal{M}_{\mathcal{S}'}$ is referred to as a matching in the subgraph $\mathcal{G}_{\mathcal{S}'} = (\mathcal{S}', \mathcal{S}', \mathcal{S}' \times \mathcal{S}')$, $w(\cdot)$ defined in (3) is the weight sum of a matching, and $\mathcal{M}_{\mathcal{S}'}^*$

is maximum weighted matching in the modified subnetwork formed by $\mathcal{S}'$ with the modified weight β_{ij}, for any $\mathcal{S}' \subseteq \mathcal{S} \subseteq \mathcal{K}$. Here the modified weight β_{ij} is given by

$$\beta_{ij} = \begin{cases} \alpha_{ij}, \, i \neq j \\ 0, \quad i = j \end{cases} \tag{18}$$

Then the optimization problem in (16a) can be reformulated as

$$\begin{aligned} \max & \left\{ \sum_{i \in \mathcal{K}} \sum_{j \in \mathcal{K}} \alpha_{ij} x\left(i,j\right) - w\left(\mathcal{M}^*_{\mathcal{K}'}\right) \right\} \\ \text{s.t.} & \quad (16b), (16c), (16d), (16e), (16f). \end{aligned} \tag{19}$$

with

$$\begin{aligned} w\left(\mathcal{M}^*_{\mathcal{K}'}\right) = \max & \sum_{i \in \mathcal{K}} \sum_{j \in \mathcal{K}} \alpha'_{ij} x\left(i,j\right) \\ \text{s.t.} & \quad (16b), (16c), (16d), (16e), (16f). \end{aligned} \tag{20}$$

where $\mathcal{M}^*_{\mathcal{K}'}$ is the matching with the maximum-weight sum in the network $\mathcal{G}_{\mathcal{K}'}$. Here the network $\mathcal{G}_{\mathcal{K}'}$ is constructed by $\mathcal{K}$ with modified weight α'_{ij}, which is given as follows

$\alpha'_{ij} =$

$$\begin{cases} \alpha_{ij} \; i,j \in \mathcal{N}_k & \text{(21a)} \\ \alpha_{ij} \text{ if} \left(i,j\right) = 0, \; \underset{q \neq j}{\exists x} \left(i,q\right) = 1, i \in \mathcal{N}_k, \; j,q \in \mathcal{N}^C & \text{(21b)} \\ \alpha_{ij} \text{ if } x\left(i,j\right) = 0, \; \exists \underset{p \neq i}{x} \left(p,j\right) = 1, i,p \in \mathcal{N}^C, \; j \in \mathcal{N}_k & \text{(21c)} \\ \alpha_{ij} \text{ if } x\left(i,j\right) = 0, \; \exists \left(\underset{p \neq i}{x(p,j)} = 1, \; \underset{q \neq j}{x(i,q)} = 1 \right) i \in \mathcal{N}_k, \; j,q \in \mathcal{N}^C & \text{(21d)} \\ 0 \qquad \text{others} & \text{(21e)} \end{cases}$$

where $\mathcal{N}_k \triangleq \{kN_a + 1, kN_a + 2, ..., (k+1)\,N_a\}$, and $k = 0, ..., N_f - 1$. Notice that the modified weight α'_{ij} is determined after the user-AP is paired in the matching $\mathcal{M}_{\mathcal{S}}$, and here it is not restricted that the matching $\mathcal{M}_{\mathcal{S}}$ is the maximum matching. $w\left(\mathcal{M}^*_{\mathcal{K}'}\right)$ can be considered as the maximum sum-weight of the matching in an interference network after the user-AP is paired.

The corresponding matrix $\mathbf{A}'$ of the interference network can be shown in Fig. 5. It can be found in the figure, (a) represents the self-interference between APs due to full duplex transmission at the same frequency. (b) represents the interference received by the paired UE under the same frequency from other APs not paired with it. (c) represents the interference received by the AP under the same frequency from UEs paired with other APs. And (d) represents the interference among paired UEs under the same frequency. Here the determination of (a), (b), (c), and (d) is based on (21a), (21b), (21c), and (21d), respectively.

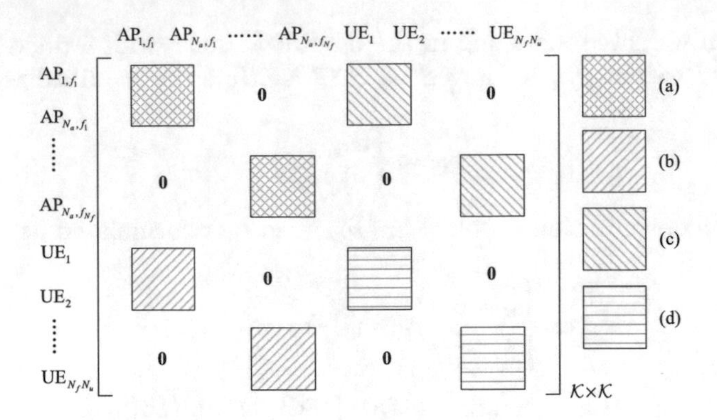

Fig. 5. The channel strength level matrix A of graph $\mathcal{G}$.

Under the constraints in (19), a subnet $\mathcal{S}$ is formed by the red edges in the bipartite graph $\mathcal{G}$ in Fig. 4. The maximum sum-weight of the matching can be denoted by

$$\begin{aligned} w\left(\mathcal{M}_{\mathcal{S}}^{*}\right) &= \max \sum_{i \in \mathcal{K}} \sum_{j \in \mathcal{K}} \alpha_{ij} x\left(i,j\right) \\ \text{s.t.} \quad & (16\text{b}), (16\text{c}), (16\text{d}), (16\text{e}), (16\text{f}). \end{aligned} \tag{22}$$

Then we can get

$$\begin{aligned} & \sum_{i \in \mathcal{K}} \sum_{j \in \mathcal{K}} \alpha_{ij} x\left(i,j\right) \leq w\left(\mathcal{M}_{\mathcal{S}}^{*}\right) \\ \text{s.t.} \quad & (16\text{b}), (16\text{c}), (16\text{d}), (16\text{e}), (16\text{f}). \end{aligned} \tag{23}$$

For further analysis, we define $w\left(\mathcal{M}_{\mathcal{K}}^{\circ}\right)$ as follows. The original network performs maximum weighted matching $\mathcal{M}_{\mathcal{S}}^{*}$ with the channel gain α_{ij} as the weight under the constraints of (16b), (16c), (16d), (16e), and (16f). The weight is then modified as α'_{ij} in (23) form the new maximum matching $\mathcal{M}_{\mathcal{K}}^{\circ}$, and the corresponding maximum-weight sum is $w\left(\mathcal{M}_{\mathcal{K}}^{\circ}\right)$.

Different from $w\left(\mathcal{M}_{\mathcal{K}'}^{*}\right)$ in (20), the maximum weighted sum matching is first performed. The relationship between $w\left(\mathcal{M}_{\mathcal{K}'}^{*}\right)$ and $w\left(\mathcal{M}_{\mathcal{K}}^{\circ}\right)$ can be that $w\left(\mathcal{M}_{\mathcal{K}'}^{*}\right) \geq w\left(\mathcal{M}_{\mathcal{K}}^{\circ}\right)$, or $w\left(\mathcal{M}_{\mathcal{K}'}^{*}\right) < w\left(\mathcal{M}_{\mathcal{K}}^{\circ}\right)$.

If $w\left(\mathcal{M}_{\mathcal{K}'}^{*}\right) \geq w\left(\mathcal{M}_{\mathcal{K}}^{\circ}\right)$, then based on (22) and (23) we can obtain that

$$\begin{aligned} & \sum_{i \in \mathcal{K}} \sum_{j \in \mathcal{K}} \alpha_{ij} x\left(i,j\right) - w\left(\mathcal{M}_{\mathcal{K}'}^{*}\right) \leq \max \left[\sum_{i \in \mathcal{K}} \sum_{j \in \mathcal{K}} \alpha_{ij} x\left(i,j\right)\right] - w\left(\mathcal{M}_{\mathcal{K}}^{\circ}\right) \\ & s.t. \quad (16\text{b}), (16\text{c}), (16\text{d}), (16\text{e}), (16\text{f}). \end{aligned} \tag{24}$$

Now we consider that $w\left(\mathcal{M}_{\mathcal{K}'}^{*}\right) < w\left(\mathcal{M}_{\mathcal{K}}^{\circ}\right)$. When $\sum_{i \in \mathcal{K}} \sum_{j \in \mathcal{K}} \alpha_{ij} x\left(i,j\right) < w\left(\mathcal{M}_{\mathcal{S}}^{*}\right)$, it means the matching $\mathcal{M}_{\mathcal{S}}$ is not the maximum matching $\mathcal{M}_{\mathcal{S}}^{*}$. The matching $\mathcal{M}_{\mathcal{K}'}^{*}$ is performed after the matching $\mathcal{M}_{\mathcal{S}}$ with the modified weight

α'_{ij}. Hence the matching $\mathcal{M}^*_{\mathcal{K}'}$ contain some edges in the matching $\mathcal{M}^*_{\mathcal{S}}$. Suppose the edge (i', j') is in both the matching $\mathcal{M}^*_{\mathcal{K}'}$ and $\mathcal{M}^*_{\mathcal{S}}$, then this edge can not be in the matching $\mathcal{M}^\circ_{\mathcal{K}}$, since the matching $\mathcal{M}^\circ_{\mathcal{K}}$ is performed after the maximum matching $\mathcal{M}^*_{\mathcal{S}}$.

Since in the full duplex system, the self-interference signal cancellation method can be employed to reduce the AP-to-AP interference signal to be smaller than the interested signal between AP-to-UE, then the corresponding weight of the AP-to-AP edge is smaller than the weight $\alpha'_{i'j'}$ of the edge (i', j'). Considering $w(\mathcal{M}^*_{\mathcal{K}'}) < w(\mathcal{M}^\circ_{\mathcal{K}})$, at least one user-to-user edge $\{(n, m) : n, m \in \mathcal{N}^c\}$ in the matching $\mathcal{M}^\circ_{\mathcal{K}}$ with weight α'_{nm} is lager than $\alpha'_{i'j'}$. Based on [6] and [7], it is found that the conditions for obtaining GDoF are not satisfied. This means the user node n or the user node m must be silent to obtain GDoF. Hence the users that do not meet the conditions for obtaining GDoF do not participate in the user-AP assignment, which means that it can only be $w(\mathcal{M}^*_{\mathcal{K}'}) \geq w(\mathcal{M}^\circ_{\mathcal{K}})$. In consequence, (24) can always be established.

According to (24), the optimization problem in (19) can be written as

$$\max\left\{\sum_{i\in\mathcal{K}}\sum_{j\in\mathcal{K}}\alpha_{ij}x(i,j) - w(\mathcal{M}^*_{\mathcal{K}'})\right\} = \max\left[\sum_{i\in\mathcal{K}}\sum_{j\in\mathcal{K}}\alpha_{ij}x(i,j)\right] - w(\mathcal{M}^\circ_{\mathcal{K}})$$
$$\text{s.t.} \quad (16\text{b}), (16\text{c}), (16\text{d}), (16\text{e}), (16\text{f}). \tag{25}$$

According to the above analysis, we can do the maximum matching through the following steps:

- `Step` 1: Achieve $w(\mathcal{M}^*_{\mathcal{S}})$, according to (22).
- `Step` 2: Based on the matching $\mathcal{M}^*_{\mathcal{S}}$, α'_{ij} in (23) is obtained to form the new subnetwork, and the corresponding maximum matching $\mathcal{M}^\circ_{\mathcal{K}}$ is determined.
- `Step` 3: Based on the GDoF conditions and $\mathcal{M}^\circ_{\mathcal{K}}$, the UE node to be silent is determined.
- `Step` 4: Remove the silent UE node in the network, and return to step 1 to perform iteratively until there are no UE nodes which need to be silent.

4 Simulation

To confirm the theoretical analysis, the performance of the proposed algorithms is evaluated by using computer simulations. We set up a small cell topology as illustrated in Fig. 6, where The users are assumed to be uniformly distributed. All simulation results are obtained by averaging over 10000 runs for different locations of users. The general simulation parameters are taken from [4,5] and listed in Table 1, where d is the distance (in kilometers) among nodes.

Firstly, the effect of RSI on the average sum rate of the network in both FD and HD modes are simulated and shown in Fig. 7, respectively, under the same conditions. It can be found that with the increase of RSI, the average sum rate of the FD network gradually decreases, and the corresponding advantage

Table 1. Simulation parameters

Parameter	Value
Carrier frequency/Bandwidth	2 GHz/10 MHz
Distance between the AP and nearest UE	>15 m
Radius of small cell (R)	100 m
Noise power spectral density	−174 dBm/Hz
Receiver noise figure	5 dB
Path loss from a AP to a UE	$103.8 + 20.9\log_{10}(d)$ dB
Path loss among UEs	$145.4 + 37.5\log_{10}(d)$ dB
Maximum transmit power at AP (P_{AP})	10 or 26 dBm
Maximum transmit power at UE (P_{UE})	0 or 10 dBm
Antenna gain	0 dB

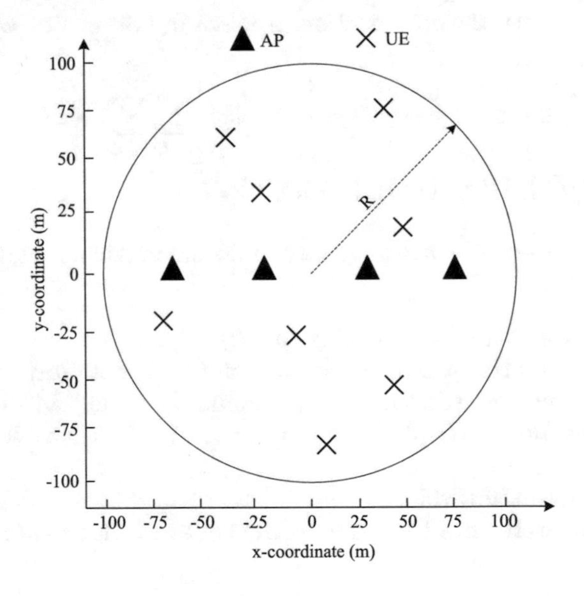

Fig. 6. A small cell topology for our simulations.

decreases compared with the HD network. Therefore, it is important to control the magnitude of RSI. Figure 7 also compares the network performance under the number of $N_a = 2N_u = 4$ and $N_a = 4N_u = 8$. It can be seen that due to the small cell radius, the increase in the number of nodes leads to an increase in the interference between UEs, thus resulting in a decrease in the network average sum rate. In addition, the network average sum rate under the two cases of high transmit power ($P_{AP} = 26$ dBm, $P_{UE} = 10$ dBm) and low transmit power ($P_{AP} = 10$ dBm, $P_{UE} = 0$ dBm) are also simulated and compared. It can be seen that

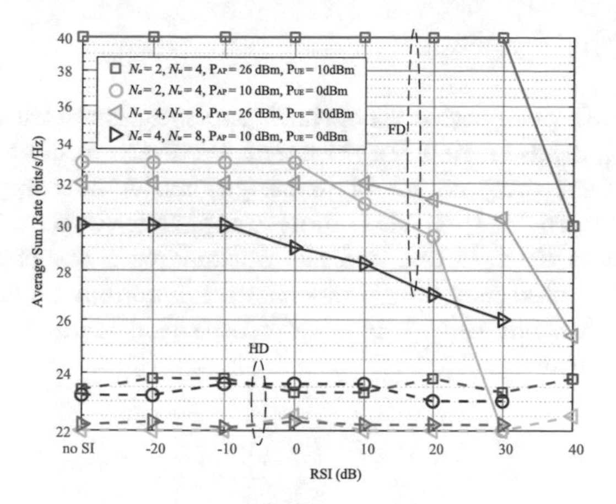

Fig. 7. Average sum rate versus RSI.

the average sum rate of high transmit power is higher than that of low transmit power, because the signal-to-noise ratio (SNR) of high transmit power is higher than that of low transmit power.

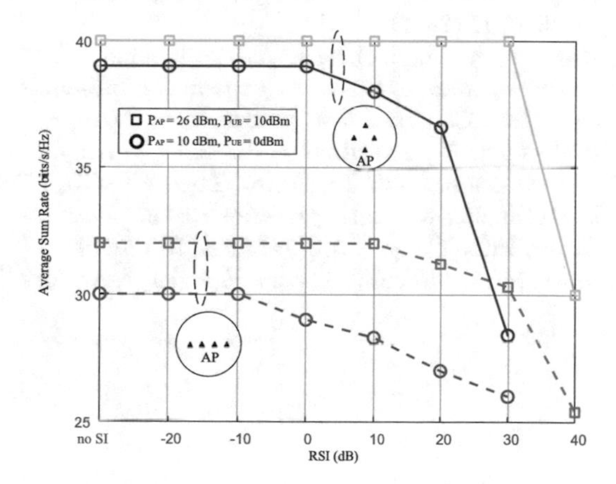

Fig. 8. The effect of locations of APs.

Figure 8 shows the effect on the average sum rate of AP location placement within a cell. The average sum rates of 4 APs located at coordinates $\{(-50, 0), (-20, 0), (20, 0), (50, 0)\}$ and $\{(-50, 0), (50, 0), (0, 50)\ (0, -50)\}$ are compared. It can be seen from the simulation results in Fig. 8 that the placement of APs has a great influence on the average sum rate of the network. Therefore, based on the algorithm proposed in this paper, the study of the optimal location layout of APs is a potential future research point.

5 Conclusion

Based on the GDoF perspective, an optimal user assignment method is provided for the FD distributed-AP multi-user system. We first formulate the optimality problem as a linear program for bipartite graphs. A new expression for the linear program is provided, which is more compact and useful than the initial optimality problem since it can find the optimal user assignment by solving two maximum weighted matching problems in polynomial time. By simulation, we have shown that our user assignment mechanism achieves significant sum throughput improvement.

References

1. Bharadia, D., McMilin, E., Katti, S.: Full-duplex radios. In: Proceedings of ACM SIGCOMM, pp. 375–386 (2013)
2. Duarte, M., Dick, C., Sabharwal, A.: Experiment-driven characterization of full-duplex wireless systems. IEEE Trans. Wirel. Commun. **11**(12), 4296–4307 (2012)
3. Geng, C., Jafar, S.A.: On the optimality of treating interference as noise: compound interference networks. IEEE Trans. Info. Theory **62**(8), 4630–4653 (2016)
4. Nguyen, V.-D., Nguyen, H.V., Nguyen, C.T., Shin, O.-S.: Spectral efficiency of full-duplex multi-user system: beamforming design, user grouping, and time allocation. IEEE Access **5**, 5785–5797 (2017)
5. Nguyen, H.V., Nguyen, V.-D., Dobre, O.A., Yongpeng, W., Shin, O.-S.: Joint antenna Array mode selection and user assignment for full-duplex MU-MISO systems. IEEE Trans. Wirel. Commun. **18**(6), 2946–2963 (2019)
6. Geng, C., Naderializadeh, N., Avestimehr, S., Jafar, S.A.: On the optimality of treating interference as noise. IEEE Trans. Inf. Theory **61**(4), 1753–1767 (2015)
7. Yi, X., Caire, G.: Optimality of treating interference as noise: a combinatorial perspective. IEEE Trans. Info. Theory **62**(8), 4654–4673 (2016)
8. Lovász, L., Plummer, M.D.: Matching Theory. Elsevier, New York, NY, USA (1986)

CDN Service Detection Method Based on Machine Learning

Yijing Wang[1,2](✉) and Han Wang[1]

[1] Institute of Information Engineering, Chinese Academy of Sciences, Beijing, China
{wangyijing,wanghan8419}@iie.ac.cn
[2] School of Cyber Security, University of Chinese Academy of Sciences, Beijing, China

Abstract. Content Delivery Network (CDN) is an emerging network acceleration technology and an important infrastructure on the Internet. However, there are currently a large number of domestic domain names that use CDN services in confusion, setting up numerous obstacles to the management of domestic basic resources, which in turn leads to potential security risks for the network security of the entire country. Currently, CDN domain names are detected mainly by using the character features of CDN domain names, HTTP keywords, and DNS records, and the recognition range is very limited. In response to this problem, this article introduces the basic principles and workflow of CDN in detail, analyzes the characteristics and related attributes of CDN domain names, and uses random forest classification algorithms to establish a CDN service detection model based on machine learning., It mainly detects CDN domain names and CDN acceleration nodes, and verifies the accuracy of the proposed detection method through experimental analysis, and explains the necessity of CDN service detection.

Keywords: Domain name system · CDN · Machine learning

1 Introduction

With the rapid growth of Internet users, the image, audio, video and other services provided by content providers account for an increasing proportion of the Internet. The expansion of user scale leads to the extension of the average distance of network access, resulting in a rapid increase in network load, which seriously affects the quality of user access.

When users use the Internet, the demand for the browsing speed and effect of the website is gradually increasing, and there are certain requirements for the access speed. Traditional caching technology often only caches relatively small files such as static files and pictures, but cannot do much for large files with strong interaction. In addition, when the bandwidth between users and websites is blocked, poor access quality is an urgent problem for Internet users in other regions. The emergence of CDN technology effectively solves these problems.

The basic idea of CDN is to avoid bottlenecks and links on the Internet that may affect the speed and stability of data transmission, so as to make content transmission

© The Author(s), under exclusive license to Springer Nature Switzerland AG 2023
L. C. Tang and H. Wang (Eds.): BDET 2022, LNDECT 150, pp. 161–174, 2023.
https://doi.org/10.1007/978-3-031-17548-0_15

faster and more stable. By placing acceleration node servers everywhere in the network to form a layer of intelligent virtual network on the basis of the existing Internet, the CDN system can real-time based on network traffic and the connection and load status of each node, as well as the distance and response to the user Comprehensive information such as time redirects the user's request to the service node closest to the user. Its purpose is to enable users to obtain the desired content nearby, to solve the situation of Internet network congestion, and to improve the response speed of users accessing websites.

At present, thousands of servers deployed by global CDN service providers around the world constitute a key part of the Internet infrastructure while providing services. The large-scale application of CDN also brings new security challenges to the cyberspace. In particular, the current situation in which domestic domain names use CDN services chaotically has set up many obstacles to the management of domestic basic resources, which in turn leads to potential security risks in the network security of the entire country. Large-scale detection of CDN services will be of great significance.

2 Current Research Status

In 2019, Nguyen et al. discovered a novel cache system poisoning attack against CDNs as shown in [3], and in 2020, Li et al. proposed an HTTP range request amplification attack against CDNs as shown in [4]. The combination of CDN domain name identification detection and defense is very significant to cyberspace mapping.

At present, the methods proposed by domestic and foreign scholars for CDN domain name identification mainly include: Huang and Adhikari and others use the features of Canonical Name Record (CNAME) to identify CDN domain names that use designated service providers as shown in [5, 6]. Guo et al. used HTTP error messages returned by the server to discover CDN nodes as shown in [7]. Timm constructs hostnames by guessing the naming rules of CDN edge servers and performs DNS probes to obtain more node IPs as shown in [8]. Although these methods are simple and effective, they can obtain the data of one or several specific CDN service providers to measure the scale and performance of the CDN, but their identification scope is very limited and can only identify some specific CDN domain names.

Aiming at the difficulty in distinguishing between CDN domain names and Fast-Flux (FF) domain names, Using Long Short-Term Memory (LSTM) network, Chen distinguished between FF domain names and CDN domain names based on domain name characteristics, empirical information, geographic and time-related characteristics as shown in [9]. Li et al. identified CDN domain names based on machine learning based on the related features of domain name system records as shown in [10], but they required multi-region and multi-node data acquisition for the target, and manually identified a large number of CDN domain names to construct a sample set. Therefore, how to identify large-scale CDN domain names at a low cost is still an urgent problem to be solved.

3 Implementation Principle of CDN

3.1 Technical Overview of CDN

A CDN provides a mechanism. When a user requests content, the requested content can be provided to the user by the Cache that is delivered at the fastest speed. The process of selecting the "optimal" is load balancing. From a functional point of view, a typical CDN system consists of a distribution service system, a load balancing system and an operation management system.

Distribution Service System. The most basic work unit of the distribution service system is the Cache device. The cache (edge cache) is responsible for directly responding to the end user's access request and quickly providing the locally cached content to the user. At the same time, the cache is also responsible for synchronizing content with the source site, acquiring updated content and content not available locally from the source site and saving it locally. The number, scale, and total service capability of Cache devices are the most basic indicators to measure the service capability of a CDN system.

Load Balancing System. The main function of the load balancing system is to schedule access to all users who initiate service requests, and to determine the final actual access address provided to users. The two-level scheduling system is divided into Global Server Load Balance (GSLB) and Server Load Balance (SLB). GSLB mainly determines the physical location of the cache that provides services to users by making an "optimal" judgment on each service node based on the principle of user proximity. SLB is mainly responsible for device load balancing within the node.

Operation Management System. The operation management system is divided into operation management and network management subsystems, which are responsible for the collection, sorting, and delivery work necessary for interaction with external systems at the business level, and have functions such as customer management, product management, billing management, and statistical analysis.

3.2 Principles of CDN Technology

CDN is to add a new network architecture to the existing network. CDN publishes and transmits the content of the source site to the area closest to the user, so that the user can access the desired content nearby, thereby improving the response speed of the user's visit. The basic principle of CDN is to rely on the acceleration nodes placed in various places as cache servers, and through functional modules such as global scheduling and content distribution, the part of the content that users need is deployed to the location closest to the user, so as to improve the Internet network congestion problem and improve response speed of website.

The biggest difference between CDN services and traditional network services is the access method. Traditionally, all user requests for the same content are concentrated on the same target server. After the introduction of CDN acceleration, the resolution right of the user's request is handed over to the CDN scheduling system, and then the user request is directed to the CDN acceleration node with the best performance closest to the user, and finally the node provides services for the user.

CDN user request scheduling is divided into global scheduling and local scheduling. Global scheduling is to analyze and decide between nodes according to the geographical location of the user, and transfer the user request to the acceleration node closest to the user. Local scheduling is limited to a certain range, and it is more inclined to pay attention to the specific health status and load of CDN server equipment. According to the real-time response time, users' tasks are allocated to the most suitable server for processing, and more fine-grained scheduling is performed. Decision-making, realize real intelligent communication, give full play to the best performance of the server cluster, and effectively solve the problem of excessive system load caused by excessive user access requests.

There are three ways of content distribution: Push, Pull and mixed distribution. Push is an active distribution method, initiated by the CDN content management system to actively distribute content from the source site or central content library to each CDN edge acceleration node. Push distribution is an intelligent distribution strategy, which intelligently decides whether to actively distribute content according to user access statistics and preset content distribution rules; Pull is a passive distribution method, driven by user requests. When the content requested by the user is accelerated on the local edge CDN node When the CDN node does not exist, the CDN node starts the Pull mode to pull the content in real time from the content source or other CDN nodes, and in the Pull mode, the content is distributed on demand; the hybrid distribution method is the combination of Push distribution methods and Pull distribution methods. There are various schemes for hybrid distribution. The most common hybrid distribution mechanism is to use the Push method to pre-push content, and then use the Pull method to pull. The hybrid distribution method can dynamically adjust the distribution of content in the content distribution system in a push-pull manner according to the current content service status in the content distribution system.

CDN is a combined technology, including the source sites, cache servers, acceleration nodes, and user clients. Figure 1 shows the global structure of CDN.

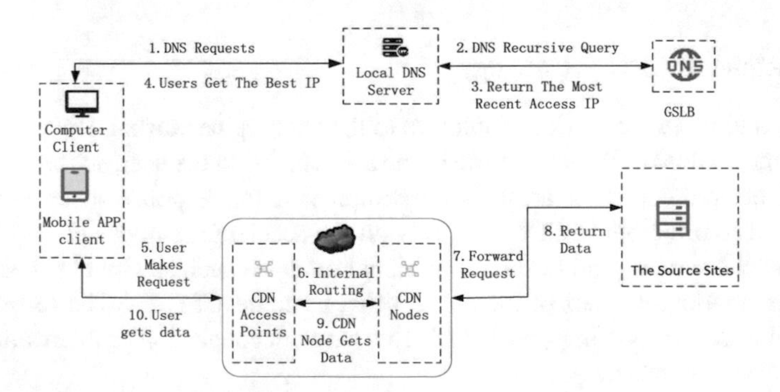

Fig. 1. Global structure diagram of CDN.

3.3 Workflow of CDN

The workflow of CDN is as follows:

1. The user clicks the content URL on the website page, and after the local DNS system analysis, the DNS system will finally give the domain name resolution right to the CDN dedicated DNS server pointed to by CNAME.
2. The DNS server of the CDN returns the IP address of the CDN's global server load balancing device to the user.
3. The user initiates a content URL access request to the CDN's global server load balancing device.
4. The CDN global server load balancing device selects a regional server load balancing device in the region to which the user belongs according to the user's IP address and the content URL requested by the user, and tells the user to initiate a request to this device.
5. After comprehensive analysis based on the following conditions, the regional server load balancing device will return the IP address of a cache server to the global server load balancing device.

- According to the user's IP address, determine which server is closest to the user.
- According to the content name carried in the URL requested by the user, determine which server has the content required by the user.
- Query the current load of each server to determine which server still has service capabilities.

6. The global server load balancing device returns the IP address of the server to the user.
7. The user initiates a request to the cache server, and the cache server responds to the user request and transmits the content required by the user to the user terminal. If there is no content that the user wants on the cache server, and the regional server load balancing device still assigns it to the user, then the server will request the content from its upper-level cache server until it can be traced back to the origin server of the website Pull content locally.

The DNS server resolves the domain name into the cache server IP address of the corresponding node according to the user's IP address, so as to realize the user's nearest access. Websites using CDN only need to hand over their domain name resolution rights to the CDN's Global Server load balancing (GSLB) device and inject the content to be distributed into the CDN to achieve content acceleration.

4 CDN Domain Name Detection Method Based on Machine Learning

4.1 Detection of CDN Domain Names

Three major operators have been deployed in 31 provinces and municipalities directly under the central government, and a total of 93 servers have been deployed. According to the characteristics of CDN services, domain names with widely distributed users and more frequent visits are more likely to open CDN services.

Therefore, this paper selects the top 100,000 domain names of Alexa for detection, saves the CNAME record value, summarizes the detection results of all detection servers, and preprocesses the data such as deduplication and formatting as the training set.

The system deployment diagram of domain name detection is shown in Fig. 2.

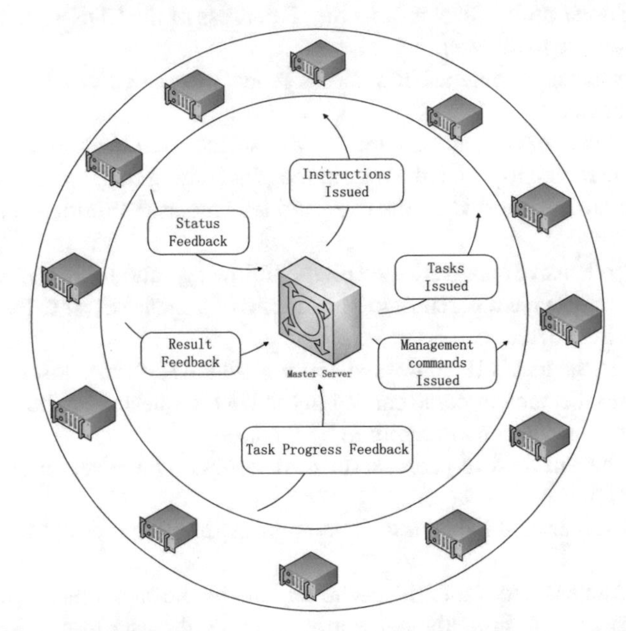

Fig. 2. Deployment diagram of CDN domain name detection system.

4.2 Analysis of CDN Domain Names

Through the analysis of the detection result data and CDN enterprise feature string data, most of the CDN domain names contain special strings such as "cdn" and "dns", which are combined with the CDN domain name and its corresponding attributes such as the source site unit information of the acceleration node as the CDN domain name feature. The CDN domain name attributes are shown in Fig. 3.

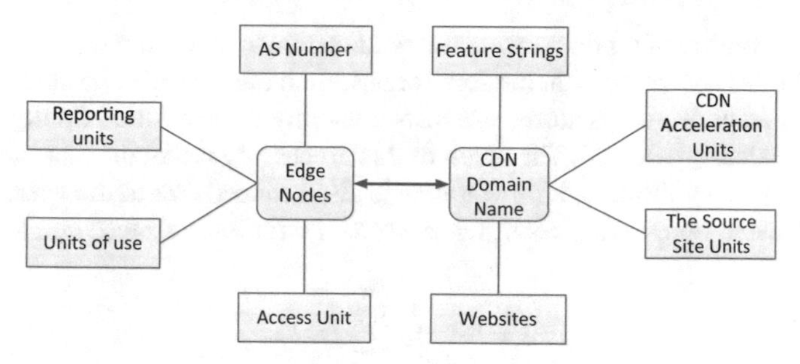

Fig. 3. Analysis of CDN service attributes.

In addition to the above attributes, take CDN service domain name as the central node, and collect statistics on the corresponding edge acceleration nodes and CDN acceleration domain names to obtain the number of edge nodes of the same service domain name and the number of corresponding acceleration domain names. These two statistical results are also important features of CDN domain name detection.

4.3 Bagging Algorithm

Bagging (Bootstrap aggregating), also known as bagging algorithm, is a group learning algorithm in the field of machine learning. It is an ensemble method used to improve the accuracy of weak classifiers. The basic idea of this method is to re-sample the training set with random samples with replacement to form multiple training subsets that are similar in size to the training set but different from each other., on this basis, multiple base classifiers are formed, and the majority voting principle is used on the basis of multiple base classifiers. It uses a randomly selected subset of the training set to train each model, which helps reduce variance and helps avoid overfitting. Its algorithm process is as follows:

1. Extract the training set from the original sample set. Each round uses the Bootstraping (with replacement) method to extract n training samples from the original sample set (in the training set, some samples may be drawn multiple times, and some samples may not be drawn once). A total of k rounds of extraction are performed to obtain k training sets. (We assume here that the k training sets are independent of each other, in fact not completely independent).
2. Each time a training set is used to obtain a model, and k training sets are used to obtain a total of k models. But it's the same model. (Note: Although the k training sets overlap and are not completely independent, the trained models are not completely independent because they are the same model. There is no specific classification algorithm or regression method here. We can use different classification or regression methods according to specific problems. Regression methods such as decision trees, perceptrons, etc.).
3. For classification problems: vote the k models obtained in the previous step to get the classification results; for regression problems, calculate the mean of the above models as the final result. (Equal importance for all models).

Use the algorithm of random forest and random tree to randomly select some features of the training set for training. In random forests, each tree model is trained with bagged sampling. In addition, the features are also randomly selected, and finally the trained tree is also randomly selected. The result of this processing is that the bias of the random forest increases very little, and the variance is also reduced due to the averaging of the weakly correlated tree model, resulting in a model with less variance and less bias.

$$f(x) = \frac{1}{M}\sum_{m=1}^{M} f_m(x)$$

In an extreme random tree algorithm, the random application is more thorough: the threshold for dividing the training set is also random, that is, the training set obtained from each division is different, which usually reduces the variance further, but at the cost of a slight increase in bias.

ALGORITHM 1: Pseudo code flow of Bagging algorithm

Enter training: $D=\{(x_1, y_1), (x_2, y_2), \ldots, (x_m, y_m)\}$

Basic Learning Algorithms: ב

Number of training rounds: T

process:

For t=1, 2, …, T do

$\boldsymbol{h_t} = ב(D, D_{bs})$ (D_{bs} is the sample distribution generated by autonomous sampling)

End for

output: $H(x) = \arg max_{y\in\gamma} \sum_{t=1}^{T} I(h_t(x) = y)$

The flowchart is shown in Fig. 4.

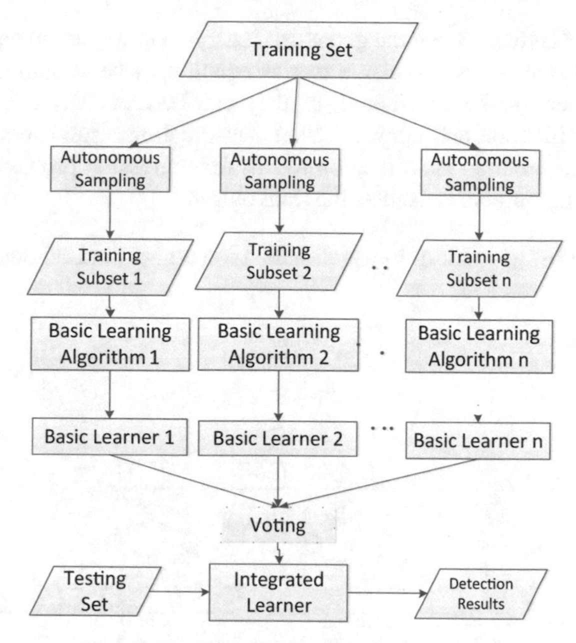

Fig. 4. Flowchart of the Bagging algorithm.

The advantage of bagging is that when there is noisy data in the original sample, through bagging sampling, then 1/3 of the noisy samples will not be trained. For classifiers that are affected by noise, bagging is helpful for the model. Therefore, bagging can reduce the variance of the model, is not easily affected by noise, and is widely used in unstable models or models that tend to overfit.

This paper uses random forest, that is, a combination of Bagging and decision tree, to detect CDN domain names. A forest composed of many independent decision trees, each tree is independent of each other, so when the model is combined, the weight of each tree is equal, that is, the final classification and detection result is determined by voting.

Selection of Sample Set. Assuming that the original CDN domain name sample set has a total of N domain names, each round extracts N domain names from the original sample set by Bootstraping (with replacement sampling) to obtain a training set of size N. In the process of extracting the original sample set, there may be domain names that are repeatedly extracted, or there may be domain names that have not been extracted once. A total of k rounds of extraction are performed, and the training sets extracted in each round are $T_1, T_2, \ldots T_k$ respectively.

Generation of Decision Tree. The feature space of the CDN domain name has N features. In the process of generating the decision tree in each round, n features ($n < N$) are randomly selected from the N features to form a new feature set. Feature set to generate a decision tree. A total of k decision trees are generated in k rounds, because the selection of training set and feature selection of these k decision trees are random, because these k decision trees are independent of each other.

Combination of Models. Since the generated k decision trees are independent of each other, the importance of each decision tree is equal, so when combining them, their weights do not need to be considered, or they can be considered to have the same weights. For classification problems, the final classification result uses all decision tree votes to determine the final classification result; for regression problems, the mean of all decision-making outputs is used as the final output.

The flowchart of the random forest classification algorithm is shown in Fig. 5:

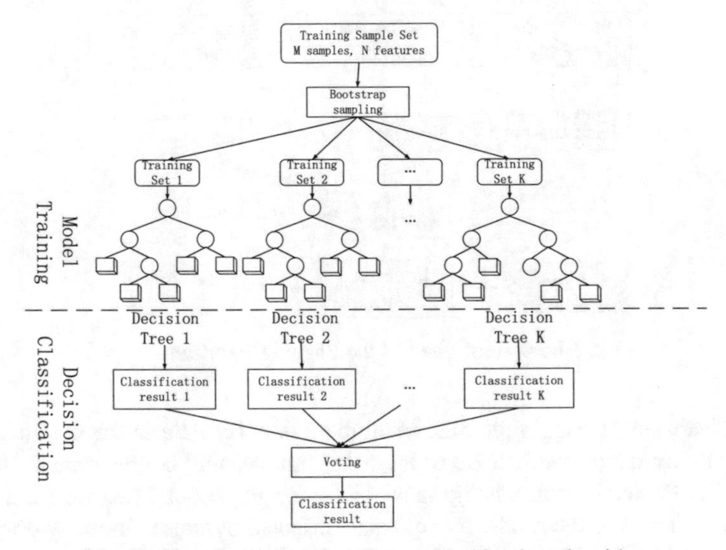

Fig. 5. Flowchart of random forest classification algorithm.

The CDN domain name detected by the classification algorithm is detected once, so as to obtain all CDN acceleration nodes.

5 Experimental Results and Analysis

5.1 Experiment Analysis

This paper detects the top 200,000 active domain names in China, extracts their CNAME records and their corresponding edge acceleration nodes, and then combines them with cdn feature strings as a training set and builds a model. Some CDN enterprise feature strings are shown in Table 1.

Table 1. Feature strings of known CDN enterprises.

CDN enterprise	Feature strings
Wangsu Technology Co., Ltd	wsglb0.com wsdvs.com cdntip.com cdn20.com wscdns.com cdn30.com
Beijing Sankuai Cloud Computing Co., Ltd	meituan.net -mtyuncdn.com
Beijing Kuaiwang Technology Co., Ltd	cloudglb.com
Hangzhou Niudun Network Technology Co., Ltd	niudunx.org
Shanghai Anchang Network Technology Co., Ltd	2x1baby.com
Shanghai Yiyun Information Technology Development Co., Ltd	scsdns.com
Shanghai Cloud Entropy Network Technology Co., Ltd	miitsp.com
Xiamen Kaoyun Co., Ltd	panshiyn.com

Some CDN enterprise acceleration nodes are shown in Table 2.

Table 2. Known enterprise CDN acceleration nodes.

CDN enterprise	IP of acceleration node
Beijing Lanxun Communication Technology Co., Ltd	58.253.70.120
Tongxing Wandian (Beijing) Network Technology Co., Ltd	115.127.226.215 115.127.226.44 14.0.108.30, 14.0.108.43
Chengdu Feishu Technology Co., Ltd	122.114.41.89, 43.230.167.13
Wangsu Technology Co., Ltd	218.60.51.254 218.60.51.9 36.33.157.53 122.138.54.174 122.143.28.61

Comparison of the Recognition Accuracy Between the CDN Domain Name Detection Model and the Missing CDN Domain Name Feature String. Compare the detection accuracy of CDN domain name feature string detection and the CDN domain name detection model proposed in this paper, as shown in Fig. 6:

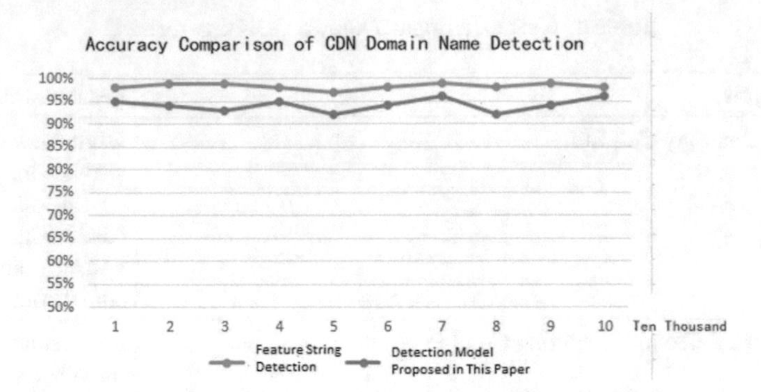

Fig. 6. The detection accuracy rate of the CDN domain name model.

It can be seen from Fig. 6 that the detection accuracy of hitting known CDN service feature strings is stable at about 98%, while most CDN domain names in traffic cannot hit known CDN service feature strings. Through the CDN domain name detection model proposed in this paper The detection of this part of CDN domain names is very close to the detection accuracy of hitting CDN fingerprints.

Comparison of the Accuracy of Direct Detection of Domain Names and Detection of CDN Domain Names to Detect Acceleration Nodes. Analyze the accuracy comparison of CDN detection for common domain names and acceleration nodes for CDN domain name detection.

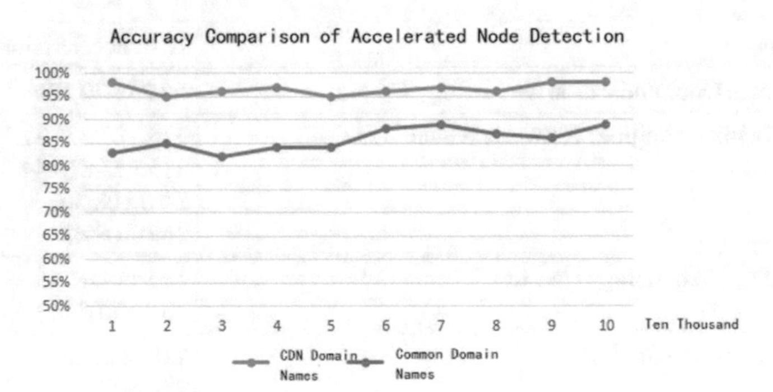

Fig. 7. Acceleration node detection accuracy comparison.

It can be seen from Fig. 7 that the accuracy of the acceleration node detected by detecting CDN domain names is significantly higher than that of detecting ordinary domain names. This is because some domain names in the network use access redirection, that is, when accessing to another alias, but it does not actually use the CDN acceleration service, and the node it finally accesses is not a CDN acceleration node. It can be seen that the accurate detection of the CDN domain name first is crucial to the detection accuracy of the entire CDN service.

5.2 Result Analysis

In order to quantitatively evaluate whether the decision tree model is capable of detecting whether the domain name test is a CDN. This paper uses the following two evaluation indicators.

1. Precision: It is the proportion of true positive samples in a certain category in the classification result that judges the correct document, which measures the accuracy of the classification system. It is for the prediction result.
2. Recall Rate: It is the proportion of how many of the original text classification results of a certain category are predicted to be correct, which measures the recall rate of the classification system. It is for the original sample

The CDN service includes domain names and acceleration nodes, and Precision of the detection and recall rate are calculated respectively. The results are shown in Table 3:

Table 3. Accuracy rate and recall rate of CDN domain name service model.

The CDN service	Precision (%)	Recall (%)
Domain names	96.4	95.8
Acceleration nodes	95.8	93.3

It can be seen from Table 3 that the CDN service detection method based on machine learning proposed in this paper has a high detection accuracy for CDN services.

6 Conclusion

This paper analyzes the basic characteristics and related attributes of CDN domain names, and uses the random forest classification algorithm to propose a CDN domain name detection model based on machine learning, and detects CDN acceleration nodes by means of detection, the experiment proves that it has a better detection effect of CDN services.

CDN service detection plays a very important role in the field of basic resources, which is of great significance to domain name IP control. Especially in the current situation of domestic domain names using CDN services in confusion, targeted investigation of CDN domain names can improve the supervision of CDN manufacturers, thereby reducing the potential risks of national cyberspace security. CDN service analysis is necessary for the development of network security.

References

1. Li, C., Wang, R., Liang, X.: Research and design of CDN technology. Internet Things Technol. (12), 28–830 (2016)

2. Yan, Z., Liu, J., Guo, H., Guo, B.: CDN domain name recognition technology based on domain name system knowledge graph. Comput. Eng. Appl. 02 (2021)
3. Nguyen, H.V., Iacono, L.L., Federrath, H.: Your cache has fallen: cache-poisoned denial-of-service attack. In: Proceedings of the 2019 ACM SIGSAC Conference on Computer and Communications Security, pp.1915–1936 (2019)
4. Li, W., Shen, K., Guo, R., et al.: CDN backfired: amplification attacks based on http range requests. In: 2020 50th Annual IEEE/IFIP International Conference on Dependable Systems and Networks (DSN), pp. 14–25. IEEE (2020)
5. Huang, C., Wang, A., Li, J., et al.: Measuring and evaluating large-scale CDNs. In: ACM IMC, vol. 8, pp. 15–29 (2008)
6. Adhikari, V.K., Guo, Y., Hao, F., et al.: Measurement study of Netflix, Hulu, and a tale of three CDNs. IEEE/ACM Trans. Netw. **23**(6), 1984–1997 (2014)
7. Guo, R., Chen, J., Liu, B., et al.: Abusing CDNs for fun and profit: security issues in CDNs origin validation. In: 2018 IEEE 37th Symposium on Reliable Distributed Systems (SRDS), pp. 1–10. IEEE (2018)
8. Böttger, T., Cuadrado, F., Tyson, G., et al.: Open connect everywhere: a glimpse at the internet ecosystem through the lens of the Netfix CDN. ACM SIGCOMM Comput. Commun. Rev. **48**(1), 28–34 (2018)
9. Chen, X., Li, G., Zhang, Y., et al.: A deep learning based fast-flux and CDN domain names recognition method. In: Proceedings of the 2019 2nd International Conference on Information Science and Systems, pp. 54–59 (2019)
10. Li, H., He, L., Zhang, H., et al.: CDN-hosted domain detection with supervised machine learning through DNS records. In: Proceedings of the 2020 the 3rd International Conference on Information Science and System, pp. 144–149 (2020)
11. Xiong, M.: Research on CDN technology and its application in broadband. Tianjin University (2015)
12. Jiang, J.: The key technology of CDN system. Digit. Commun. World (08), 12–13 (2018)
13. Tian, G.: Research on deployment modeling and deployment plan of mobile content distribution network nodes. J. Beijing Univ. Posts Telecommun. (2013)
14. Zhang, G.: Research on CDN-based video network architecture. Comput. Program. Skills Maint. (12), 161–163 (2018)
15. Tang, H., Chen, G., Chen, B., Yu, Y.: Principle and practice of content distribution network. Telecommun. Sci. **34**(11), 181 (2018)
16. Lang, F.: CDN technology and development trend analysis. Electron. World (14), 106 (2019)
17. Wang, H., Zhao, J., Han, Z., Wang, S.: Challenges brought by distributed CDN and solutions. Shandong Commun. Technol. **38**(01), 22–25 (2018)
18. Lv, H., Feng, Q.: Summary of random forest algorithm research. J. Hebei Acad. Sci. (03), 3741 (2019)
19. Conran, M.: Edge computing will push CDN into a new era. Computer World 2019-08-19(005)

A Decision Method for Missile Target Allocation

Yanan Wu and Jian Zhang(✉)

Wuhan Digital Engineering Institute, Wuhan, China
1893664@qq.com

Abstract. Aiming at the weapon target allocation problem of multi-type anti-ship missile attacking ship formations, a multi-stage multi-target optimization model is established. The NSGA-II algorithm is used to solve the model. For the results of the multiple weapon target allocation schemes generated by the NSGA-II algorithm, a new method is proposed to select. A Combinative Distance-based Assessment and Gray rational Analysis (CODAS_GRA) method based on the improved expert method is used to select the unique optimal solution from the Pareto optimal solution set generated by the NSGA-II algorithm. Experiments show that the solution selected by this method can simultaneously maximize damage to the enemy and minimize the consumption of ammunition. This method is reasonable and can provide a better choice for the commander.

Keywords: Multi-stage weapon target allocation · Multi-objective optimization · Expert weighting · CODAS method

1 Introduction

The Weapon Target Assignment (WTA) problem originated in the 1950s and is also known as the Missile Assignment Problem (MAP) [1]. The WTA problem is the key problem to be solved by the combat command and control system. The quality of the weapon target allocation scheme will directly affect the effect of actual combat.

At present, there have been a lot of research results on the WTA problem, but there are still some deficiencies. In terms of problem modeling, the existing results mostly study the static WTA problem [2–4], and there are few studies on the dynamic WTA problem considering the time dimension. From the perspective of solving methods, it can be mainly divided into three categories:

(1) Machine learning methods represented by reinforcement learning [5, 6] and neural network [7]. Reinforcement learning has the advantages of strong generalization ability [8], but when the number of weapons and targets increase, the action dimension increases significantly. How to improve the convergence of the algorithm is still a problem worthy of further study. Methods based on neural networks require a large amount of training sample data, but real battlefield data are often difficult to obtain. In general, such methods are still in their infancy and need to be further developed.

© The Author(s), under exclusive license to Springer Nature Switzerland AG 2023
L. C. Tang and H. Wang (Eds.): BDET 2022, LNDECT 150, pp. 175–191, 2023.
https://doi.org/10.1007/978-3-031-17548-0_16

(2) Heuristic optimization algorithms represented by genetic algorithms [9–11], and hybrid optimization algorithms composed of these algorithms [12, 13]. These methods are the mainstream solutions to WTA problems at present. According to the number of optimization objectives, it can be divided into single-objective optimization methods [10, 12, 13] and multi-objective optimization methods [9, 11, 14]. The result of the multi-objective optimization algorithm is the Pareto optimal solution set. How to select the unique optimal WTA scheme from the solution set is rarely mentioned in the previous work, which is the focus of this paper. The multi-criteria decision-making method can be used to select the unique solution from the Pareto optimal solution set, but the existing research results are mostly concentrated in some other fields. Wang et al. [15] summarized 10 commonly used multi-criteria decision-making methods. They applied these methods to the selection of Pareto optimal solutions for mathematical benchmark problems and chemical engineering problems, and compared the decision-making results. Wang et al. [16] added 4 other multi-criteria decision-making methods and 8 weight selection methods based on the reference [15], which were also applied to chemical engineering problems. Due to the difference of domain knowledge, there is currently no general multi-criteria decision-making method applicable to all Pareto optimal solution selection problems. For the WTA problem, the current research has achieved few results in the selection of the optimal solution. We urgently need a practical decision-making method to help commanders make decisions and select the only optimal solution from the results of multiple allocation schemes.

(3) A method for combining optimization algorithms with machine learning. Long Teng et al. [17] used the DDE algorithm to generate the training sample set of the neural network, which made the output accuracy of the neural network highly depend on the solution result of the DDE algorithm. This kind of method currently has few research results on the WTA problem, mainly focusing on some other fields [18–20].

This paper studies the background of multi-type anti-ship missiles attacking ship formations in the naval battlefield environment. Firstly, a multi-stage multi-objective mathematical model considering the time dimension is established for the dynamic WTA problem, and an elite non-dominated sorting genetic (NSGA-II) algorithm is used to solve the model. Aiming at the Pareto optimal solution set generated by the algorithm, this paper proposes a decision-making method for selecting unique optimal allocation plan.

2 Problem Modeling

This paper studies the dynamic WTA problem and adopts the "shoot-look-shoot" strategy for analysis. The "look-shoot" process is similar to the static WTA, and the dynamic WTA can be regarded as the repetition of the static WTA [21], as shown in the following formula (1) shown. The problem of weapon target allocation in each stage is similar to the first stage, the only difference is that the number of currently surviving enemy targets in each stage is different, and the number of remaining distributable weapons is different. The same solution method as the initial stage can be used.

$$DWTA = \left\{ SWTA^{(1)}, SWTA^{(2)}, \ldots, SWTA^{(T)} \right\} \tag{1}$$

For the dynamic WTA problem, the mathematical model is established as follows. Assumptions: (1) We have W different types of missiles, each type of missile has a certain quantity limit and a certain unit economic cost; the enemy fleet has a total of T ship targets, and the threat level of each target at different stages is estimated by the battlefield. (2) The same type of anti-ship missiles can be launched by different combat platforms, and each enemy ship target can be attacked by various types of anti-ship missiles. (3) For enemy ships within the range, anti-ship missiles can reach the established target through route planning and cause certain damage to the target.

1) Objective function
a) Optimization function 1: maximize damage to the enemy

$$\max f_1\left(X^t\right) = \sum_{j=1}^{T(t)} V_j(t) \left(1 - \prod_{i=1}^{W(t)} \left(1 - p_{ij}(t)\right)^{x_{ij}(t)} \right) \tag{2}$$

t indicates the current attack stage number, $X^t = \left[x_{ij}(t)\right]_{W \times T}$ represents the missile target allocation scheme in stage t, that is, the number of missiles allocated to target j by i-th type missiles in stage t. $x_{ij}(t) \in Z_0^+$, Z_0^+ represents a non-negative integer. $P^t = \left[p_{ij}(t)\right]_{W \times T}$ represents the damage probability matrix in the t stage, that is, the damage probability of i-th type missiles to the enemy target in the t stage. $V_j(t)$ represents the threat value of the jth target at stage t. $W(t)$ and $T(t)$ represent the number of remaining weapon types and the number of remaining targets in stage t respectively, $W(1) = W, T(1) = T$.

b) Optimization function 2: Minimize ammo consumption

$$\min f_2\left(X^t\right) = \sum_{j=1}^{T(t)} \sum_{i=1}^{W(t)} C_i x_{ij}(t) \tag{3}$$

C_i represents the unit economic cost of the i-th type missile. In the above model, parameters such as $p_{ij}(t)$ and $V_j(t)$ need to be determined when the battlefield information perception data is collected.

2) Constraints

$$\sum_{t=1}^{S}\sum_{i=1}^{W} x_{ij}(t) \le m_j, \forall j \in I_j \tag{4}$$

$$\sum_{t=1}^{S}\sum_{i=1}^{T} x_{ij}(t) \le n_i, \forall i \in I_i \tag{5}$$

$$0 \le \sum_{j=1}^{T} x_{ij} \le N_i^t, \forall \mathrm{i} \in I_i, \forall t \in I_t \tag{6}$$

$$\begin{cases} x_{ij} \ge 0 \quad\quad if\left(D_{\mathrm{imin}} < dis_{i,j}(t) < D_{\mathrm{imax}}\right) \\ x_{ij} = 0 \ if\left(dis_{i,j}(t) \le D_{\mathrm{imin}} \text{ or } dis_{i,j}(t) \ge D_{\mathrm{imax}}\right) \end{cases} \tag{7}$$

Constraint (4) limits the maximum amount of ammunition that can be allocated to each target during the entire combat process to avoid ammunition waste. Constraint (5) stipulates that the sum of ammunition allocated to all targets must not exceed the total amount of missiles of this type. Constraint (6) means that the limit of the number of fire channels, the maximum number of ammunition that can be launched by each type of missile in one wave. Constraint (7) represents the missile action distance constraint; $D_{i\,\min} \sim D_{i\,\max}$ represents the range of action of i-th type missiles; $dis_{ij}(t)$ represents the distance between the launch position of the i-th type missile and the target j.

3 Optimal Solution of Multi-objective WTA Scheme

The NSGA-II algorithm [22] introduces an elite strategy, which has strong optimization ability and fast running speed. The algorithm is used to solve the model, and the adaptive design is as follows.

1) **Coding design.** Select each value in the decision matrix $x_{ij}(t)$ as the chromosome, as shown in Fig. 1 below, using decimal coding, each chromosome is composed of W different gene segments, each gene segment is composed of T different locus, and the number on the locus represents the number of ammunition allocated to the jth target by the w type of missile.
2) **Genetic operator.** A binary tournament selection operator is used. Two solutions are randomly selected from the population at a time, and the better one is selected. The two-point crossover operator is used to generate offspring individuals. The starting and ending points are randomly selected, and the values of some gene loci between the two points of the parent are exchanged. The swap mutation operator is used to improve population diversity. The specific method is to randomly select two different loci on the offspring individuals and exchange the values of the two loci.
3) **Iteration termination condition.** Set the maximum number of iterations, and the algorithm will terminate if the maximum number of iterations is exceeded.

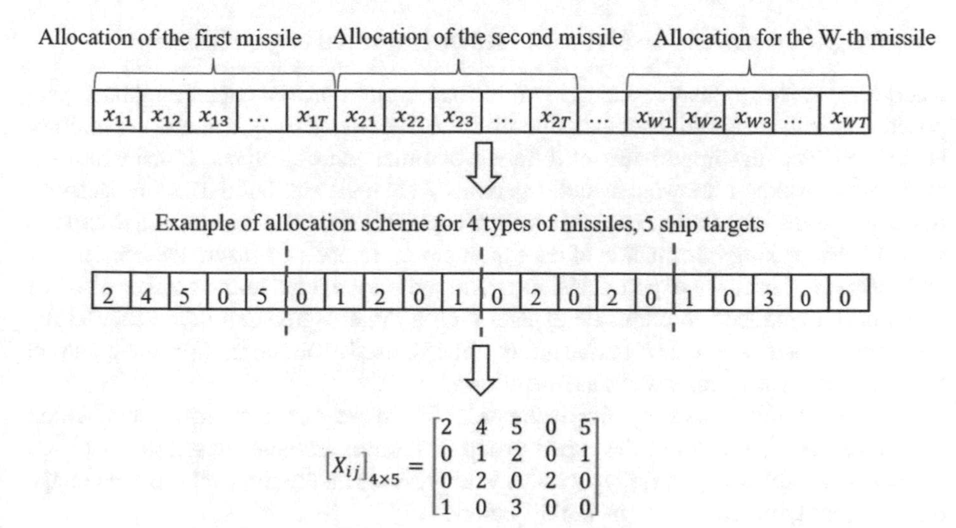

Fig. 1. Chromosome coding diagram

4 CODAS_GRA Method Based on Improved Expert Method Empowerment

The steps of the CODAS_GRA method based on the improved expert method empowerment are shown in Fig. 2 below.

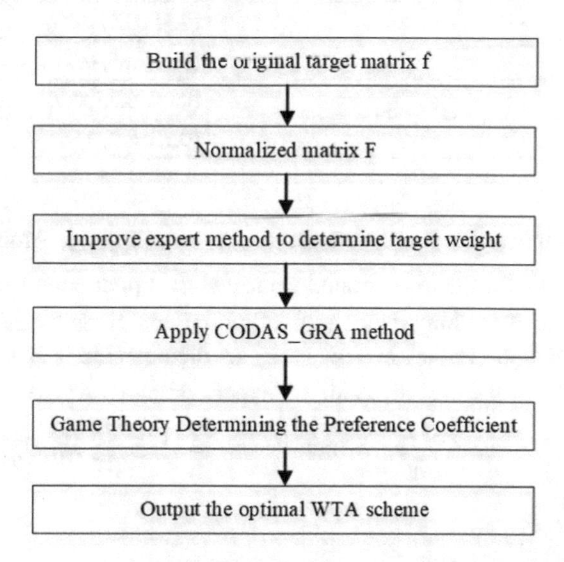

Fig. 2. Step diagram of the method

4.1 Weight Determination Method Based on Improved Expert Method

According to the application scenario of this paper, an improved expert method is proposed to calculate the weight of each optimization objective. Firstly, a number of military experts evaluate the importance of different optimization objectives. Considering the differences in expert knowledge and experience, it is believed that the expert decision-making information also has uncertainty, so triangular fuzzy numbers are used to express the decision-making information of the expert group. In order to ensure the consistency and fairness of military expert decision-making, the triangular fuzzy number distance formula is introduced to eliminate experts with a large degree of difference, and the remaining expert weight distribution information is fused to obtain the final optimization target weight. The specific steps are as follows.

Step 1: Military expert opinion expression. There are r experts and c optimization indicators. The members of the expert group give the importance description of each index, and use the triangular fuzzy number to express the qualitative evaluation language of the expert group, as shown in Table 1 below.

Table 1. Language variables of importance evaluation

Importance evaluation linguistic variables	Triangular fuzzy number
Super low	(0, 0.1, 0.25)
Very low	(0.1, 0.25, 0.35)
Low	(0.25, 0.35, 0.5)
Medium	(0.35, 0.5, 0.65)
High	(0.5, 0.65, 0.75)
Very high	(0.65, 0.75, 0.9)
Super high	(0.75, 0.9, 1)

Step 2: Determine the evaluation matrix $A = (a_{ij})_{r \times c}$. Among this, $a_{ij} = \left(a_{ij}^L, a_{ij}^M, a_{ij}^U\right)$. a_{ij}^L,a_{ij}^U are the lower bound value and the upper bound value of the triangular fuzzy number of the importance evaluation opinion given by the expert i for the index j. a_{ij}^M represent the value when the fuzzy set membership is 1.

Step 3: Calculate the fusion result of expert opinion, $A^{\circ} = (a_{F_j})_{1 \times c}$. $a_{F_j} = \left(a_{F_j}^L, a_{F_j}^M, a_{F_j}^U\right)$, $a_{F_j} = \frac{1}{r} \otimes \sum_{i=1}^{r} a_{ij}$, which is: $a_{F_j}^L = \frac{1}{r} \otimes \sum_{i=1}^{r} a_{ij}^L$, $a_{F_j}^M = \frac{1}{r} \otimes \sum_{i=1}^{r} a_{ij}^M$, $a_{F_j}^U = \frac{1}{r} \otimes \sum_{i=1}^{r} a_{ij}^U$.

Step 4: determine the dispersion matrix according to the distance formula between triangular fuzzy numbers, $D = (d_{ij})_{r \times c}$. Where d_{ij} represents the dispersion value between the decision made by each expert and the fusion result of the expert group's opinion.

Step 5: Expert opinion is eliminated. For the dispersion matrix D, count the sum of the elements of each row, and obtain the two rows with the largest and smallest sum of elements. According to this result, the corresponding two rows in the evaluation matrix A are removed, because the difference between the important decision results made by these two experts and others is the largest or smallest, so they are removed. Let the judgment matrix after removing the decision-making information of the two experts be $A' = (a'_{ij})_{(r-2) \times c}$.

Step 6: expert meaning re-integration. Perform expert opinion fusion on the A' matrix again, denoted as $\overline{A} = \left(\overline{w}_{F_j}\right)_{1 \times c}$, where $\overline{w}_{F_j} = \frac{1}{r-2} \otimes \sum_{i=1}^{r-2} a'_{ij}$.

Step 7: Calculate the defuzzification result, $r\left(\overline{w}_{F_j}\right) = \frac{1}{3}\left(\overline{w}_{F_j}^{L} + \overline{w}_{F_j}^{M} + \overline{w}_{F_j}^{U}\right)$.

Step 8: standardization processing. $w_{F_j} = \frac{r\left(\overline{w}_{F_j}\right)}{\sum_{j=1}^{c} r\left(\overline{w}_{F_j}\right)}$, Where $j = 1, 2, \ldots, c$. Then the final weight of each optimization index is $w = (w_{F_1}, w_{F_2}, w_{F_c})$.

4.2 Improved CODAS_GRA Method

The Combined Distance-based Assessment Method (CODAS) [23] uses Euclidean distance and Manhattan distance as metrics to determine the difference between each scheme and the negative ideal point, and the one with the largest difference is the recommended scheme. Human subjective cognition needs to be introduced in this calculation process. Gray Rational Analysis (GRA) obtains the recommended plan based entirely on the target matrix, and the result is more objective and accurate. But at the same time it loses the flexibility of the CODAS method. In this paper, an improved CODAS_GRA method is proposed by combining the above two methods and changing the negative ideal solution to the positive ideal solution in the CODAS method. This method can take into account the subjective flexibility of CODAS and the objective accuracy of GRA to determine the final evaluation value of each scheme. The method is implemented as the following steps. It is assumed that the objective matrix with m rows and n columns contains n optimization objective values of m schemes in the Pareto optimal solution set. f_{ij} represents the jth objective function value of the ith scheme in the objective matrix, and F_{ij} represents the result obtained after normalization, w_j represents the weight of the jth objective function.

Step 1: Construct the normalized target matrix $\left[F_{ij}\right]$.

For positive indicators: $F_{ij} = \frac{f_{ij} - \min\limits_{i \in m} f_{ij}}{\max\limits_{i \in m} f_{ij} - \min\limits_{i \in m} f_{ij}}$.

For negative indicators: $F_{ij} = \frac{\max\limits_{i \in m} f_{ij} - f_{ij}}{\max\limits_{i \in m} f_{ij} - \min\limits_{i \in m} f_{ij}}$, while $1 \le i \le m$, $1 \le j \le n$.

Step 2: Construct a normalized matrix $\left[V_{ij}\right]$ with weights. First, use the improved expert method proposed in Sect. 4.1 to determine the weight of each objective function w_j,$V_{ij} = F_{ij} \times w_j$.

Step 3: Determine the positive ideal solution $ns = \left[ns_j\right]_{1\times m}$, where $ns_j = \max_{i\in m} V_{ij}$; determine the reference vector $F_j^+ = \max_{i\in m} F_{ij}$.

Step 4: Calculate the Euclidean distance E_i and Manhattan distance T_i:

$$E_i = \sqrt{\sum\nolimits_{j=1}^{n} (ns_j - V_{ij})^2}$$

$$T_i = \sum_{j=1}^{m} |ns_j - V_{ij}|$$

Step 5: Construct the relative evaluation decision matrix R_a and the difference matrix ΔI_{ij}.

$$R_a = (h_{ik})_{m\times m}$$

$$\Delta I_{ij} = \left|F_j^+ - F_{ij}\right|$$

$$h_{ik} = (E_i - E_k) + (\psi(E_i - E_k) \times (T_i - T_k))$$

In the formula, $\psi(x) = \begin{cases} 1\ if\,|x| \ge \tau \\ 0\ if\,|x| < \tau \end{cases}$. In general, the value of τ can be 0.02.

Step 6: Calculate the score results of each scheme obtained by the CODAS method and the GRA method respectively.

$$H_i = \sum_{k=1}^{m} h_{ik}$$

$$GRC_i = \frac{1}{m}\sum_{j=1}^{n} \frac{\min\limits_{i\in m, j\in n}\left(\Delta I_{ij}\right) + \max\limits_{i\in m, j\in n}\left(\Delta I_{ij}\right)}{\Delta I_{ij} + \max\limits_{i\in m, j\in n}\left(\Delta I_{ij}\right)}$$

Step 7: Calculate the final evaluation value of each scheme. $Score_i = c_1 \times H_i + c_2 \times (1 - GRC_i)$, the scheme corresponding to the smallest $Score_i$ value is the recommended optimal WTA scheme. In this formula, the preference coefficient c_1 and c_2 can be determined by the scheme of game theory, see Sect. 4.3 for details.

4.3 Determination of Preference Coefficient Based on Game Theory

In the previous section, the CODAS method and the GRA method were combined to propose an improved CODAS_GRA method to decide the optimal weapon target allocation scheme.

It should be noted that the improved expert method proposed in Sect. 3.1 is not used here. The reason is that Sect. 3.1 determines the weights for different optimization objectives. In a naval battle, military experts with domain knowledge should decide whether to pay more attention to damage to the enemy or to save our missile resources. Therefore, subjective weighting -- improved expert method is adopted.

As can be seen from Sect. 4.2, the improved CODAS_GRA method essentially combines the advantages of CODAS and GRA. The determination of c_1 and c_2 preference coefficients is actually to weight the results of different decision-making methods. Both subjective and objective aspects should be considered comprehensively, including the preference of decision-makers towards CODAS and GRA, and the objective scores of the two methods themselves. Therefore, the idea of game theory is used to determine the preference coefficient, and the consensus or compromise of the subjective and objective weighting methods is found. The subjective weighting method adopts the analytic hierarchy process, and the objective weighting method adopts the entropy weighting method.

In the weapon target assignment problem, the multi-criteria decision-making method is used to select the unique optimal solution from the Pareto optimal set. Combined with practical application scenarios, the criteria for selecting a multi-criteria decision-making method should include: (1) The principle and algorithm of the method should be easy to understand, so that the commander can interpret the results of the scheme; (2) For convenience, the input of this method should be as little as possible; (3) The method should be easy to implement and the algorithm complexity is not high to adapt to the battlefield environment.

Game Method to Determine Preference Coefficient. The game theory comprehensive weighting method [24], which uses the game theory idea to find the Nash equilibrium point between different methods, so as to make the allocation of preference coefficients more scientific and reasonable. The specific steps are as follows.

Step 1: The weight vector u_1 determined by the AHP method and the weight vector u_2 determined by the entropy weight method are regarded as both sides of the game, and a combined weight vector $u = a * u_1^T + b * u_2^T$ is constructed, where a and b represent linear combination coefficients;

Step 2: To seek the optimal linear combination coefficient and solve the Nash equilibrium point, even if the deviation between the combination weight vector u and u_1, u_2 is the smallest, that is: $\min\left\| a \times u_1^T + b \times u_2^T - u_k^T \right\|_2, k = 1, 2$;

Step 3: From the properties of matrix differentiation, the optimal first derivative condition of the above formula is:

$$\begin{bmatrix} u_1 u_1^T & u_1 u_2^T \\ u_2 u_1^T & u_2 u_2^T \end{bmatrix} \begin{bmatrix} a \\ b \end{bmatrix} = \begin{bmatrix} u_1 u_1^T \\ u_2 u_2^T \end{bmatrix}$$

Step 4: Normalize the linear combination coefficient obtained by the above formula, namely:

$$\begin{cases} c_1 = \frac{|a|}{|a|+|b|} \\ c_2 = \frac{|b|}{|a|+|b|} \end{cases}$$

c_1 and c_2 are the final preference coefficient values obtained by the game theory method.

AHP. Analytical hierarchy process (Analytical Hierarchy Process Method, AHP) is a traditional subjective weighting method. In the problem of determining the weight of different decision-making methods, the brief application steps of this method are as follows.

Step 1: According to the selection criteria of the decision-making method, establish a hierarchical structure model as shown in Fig. 3 below.

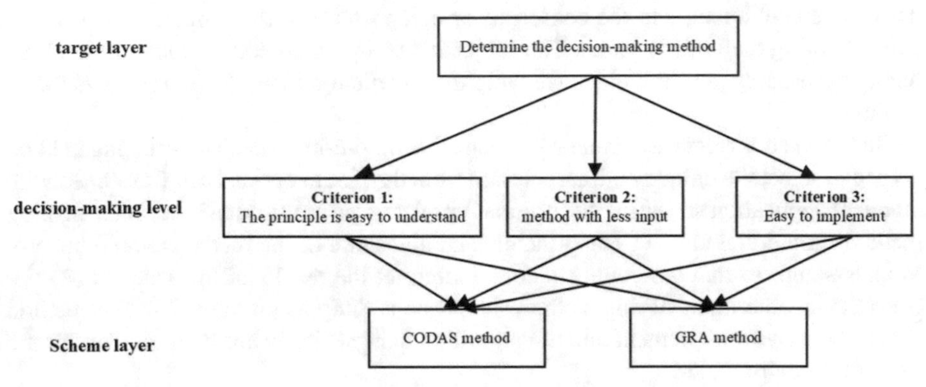

Fig. 3. Hierarchical structure

Step 2: Construct a judgment matrix according to the 1–9 scale method in combination with expert opinions.

Step 3: Perform hierarchical sorting and consistency check.

Step 4: Calculate the combined weight vector of the final solution layer to the target, which can be used as the respective weight of the CODAS method and the GRA method determined by the AHP method after normalization. It is set as the weight vector u_1.

Entropy Weight Method. As an objective weighting method, entropy weight method requires objective data for calculation. When the entropy weight method is used to determine the respective weights of the CODAS method and the GRA method, the values of the two methods under three standards need to be given first. The evaluated objects are the CODAS method and the GRA method, and the evaluation indicators are the three standards. The specific steps as follows.

Step 1: Construct the original data matrix. According to the score scale table shown in Table 2 below, the k rows q columns score matrix $\left[x_{\mathrm{ij}}\right]$ of the two methods under three standards is obtained, where x_{ij} represents the score value of the ith method under the jth criterion.

Step 2: normalize the original score data matrix to construct a normalized matrix $\left[r_{\mathrm{ij}}\right]$.

Step 3: Calculate the characteristic proportion p_{ij} of each evaluation object under each index.

$$p_{ij} = \frac{r_{ij}}{\sum_{i=1}^{k} r_{ij}}$$

Table 2. Score scale table

Scaling	Very low	Low	Medium	High	Very high
Score	0	0.25	0.5	0.75	1

Among this, $1 \leq i \leq k, 1 \leq j \leq q, 0 \leq p_{ij} \leq 10 \leq p_{ij} \leq 1$
Step 4: Calculate the entropy value of each index.

$$H_j = -\frac{1}{\ln k}\sum_{i=1}^{k} p_{ij} \ln p_{ij}$$

Step 5: Calculate the entropy weight of each index.

$$w_j = \frac{1 - H_j}{\sum_{j=1}^{q}(1 - H_j)}$$

where $0 \leq w_j \leq 1, \sum_{j=1}^{q} w_j = 1$.

Step 6: Calculate the comprehensive evaluation value of each evaluation object, and normalize it to obtain the final weight vector, which can be used as the respective weights of the CODAS method and the GRA method determined by the entropy weight method.

$$v_i = \sum\nolimits_{j=1}^{q} w_j p_{ij}$$

$$u_2 = \{v_1, v_2, ..., v_k\}$$

5 Simulation Results

5.1 Experimental Scene

At some point, we decided to attack the enemy fleet. In the initial state, we have 4 types of anti-ship missiles, and the enemy fleet has 5 ship targets. For some parameter information, please refer to Tables 3, 4, 5 and 6 below. It is stipulated that the total upper limit of ammunition allocated to each target in the whole process is 12 to avoid waste. The algorithm outputs a weapon target assignment scheme one run at a time, as a strike wave. After each wave of strikes, assess the damage degree of each target and determine the remaining ammunition amount. According to the battlefield information and Table 6, determine whether to proceed with the next wave solution. Stipulate that all targets should be severely damaged or sunk.

Table 3. Threat degree of target

Target	T1	T2	T3	T4	T5
Threat of level	0.17	0.17	0.32	0.22	0.12

Table 4. Anti-ship missile information

Types of anti-ship missiles	W1	W2	W3	W4
initial number of missiles	16	16	8	16
One wave can be launched	8	8	4	8
Unit economic cost (100 million yuan)	0.1	0.15	0.16	0.16

Table 5. Damage degree of missiles against targets

Types of anti-ship missiles	T1	T2	T3	T4	T5
W1	0.2	0.2	0.1	0.1	0.3
W2	0.3	0.3	0.2	0.2	0.2
W3	0.3	0.3	0.1	0.2	0.3
W4	0.3	0.3	0.1	0.2	0.4

Table 6. Range of damage degree

Degree of damage	Pin down	Slightly injure	Seriously injure	Sink
Interval	[0, 0.25]	(0.25, 0.65]	(0.65, 0.9]	(0.9, 1]

5.2 Experimental Results

According to the scenario in the previous section, MATLAB R2019a was used for programming, and the experiment was carried out on the host of Intel(R) Core(TM) i5-10210U CPU @ 1.60GHz 2.11 GHz, 16G memory, Windows 10. The chromosome crossover probability was 0.9, the mutation probability was 0.1. The population size was 100, and the number of iterations was 1000.

In addition, the TOPSIS method [25], the MOORA method [16], the CODAS_P method, the CODAS_N method [23], and the GRA method [15] were compared with the CODAS_GRA method proposed in this paper, where "CODAS_P" indicates that the positive ideal solution is selected during the calculation, "CODAS_N" means to select the negative ideal solution. The optimal WTA results finally selected by each method are shown in Fig. 4. The circle represents the optimal solution selected by different methods. Table 7 shows the damage value and economic cost caused by each scheme to the five targets. Table 8 shows the damage degree corresponding to the damage value and the economic cost ranking.

Table 7. Results of selected schemes by methods

Method	P1	P2	P3	P4	P5	Cost
TOPSIS	0.608	0.608	0.672	0.488	0.706	2.40
MOORA	0.906	0.918	0.824	0.866	0.894	5.84
CODAS_P	0.510	0.608	0.590	0.488	0.657	2.08
CODAS_N	0.608	0.608	0.672	0.488	0.580	2.30
GRA	0.832	0.866	0.807	0.790	0.824	4.56
CODAS_GRA	0.510	0.510	0.672	0.488	0.657	2.14

Table 8. Results of selected schemes by methods

Method	T1	T2	T3	T4	T5	Spend ranking
TOPSIS	Slightly injure	Slightly injure	Seriously injure	Slightly injure	Seriously injure	3
MOORA	Sink	Sink	Seriously injure	Seriously injure	Seriously injure	1
CODAS_P	Slightly injure	Slightly injure	Slightly injure	Slightly injure	Seriously injure	6
CODAS_N	Slightly injure	Slightly injure	Seriously injure	Slightly injure	Slightly injure	4
GRA	Seriously injure	Seriously injure	Seriously injure	Seriously injure	Seriously injure	2
CODAS_GRA	Slightly injure	Slightly injure	Seriously injure	Slightly injure	Seriously injure	5

Combined with Fig. 4 and Table 7, it can be seen that although the schemes selected by MOORA method and GRA method can cause serious injury or sinking effect to all targets, they require high economic cost and do not meet the tactical requirements of multi-wave attack. However, CODAS_P and CODAS_N have similar economic cost with CODAS_GRA, but can only cause serious injury to one target. Compared with the TOPSIS method and CODAS_GRA method, both of them can cause serious injury to T3 and T5 targets, but CODAS_GRA method has less cost and better scheme.

For the optimal scheme selected by the CODAS_GRA method, the remaining targets and the currently distributable missiles are re-determined after the execution of one wave of scheme. According to the target strike requirements, it can be known that there are three remaining targets, namely T1, T2 and T4, and the missile is surplus. The second wave of weapon target allocation scheme is continued to be solved. Assume that the effectiveness of the weapon against the target remains constant in the second stage. The results of the first-stage and second-stage weapon target allocation schemes are combined, as shown in Table 9 below.

Table 9. CODAS_GRA selected program results

	Target ship 1		Target ship 2		Target ship 3		Target ship 4		Target ship 5	
	First stage	Second stage	First stage	Second stage	First stage	Second stage	First stage	Second stage	First stage	Second stage
Missile type 1	0	1	0	1	0	0	0	1	3	0
Missile type 1	1	1	1	3	5	0	1	4	0	0
Missile type 1	1	1	1	1	0	0	2	2	0	0
Missile type 1	0	1	0	1	0	0	0	1	0	0
Final damage	0.7256		0.8655		0.6723		0.8113		0.6570	
Damage degree	Seriously injure		Seriously injure		Seriously injure		Seriously injure		Seriously injure	

As can be seen from Table 9, the first stage (first wave) plan focused its firepower on target ships 3 and 5, those with the highest and least threat, and the second stage supplemented the attack on the remaining three ships. In this paper, the CODAS_GRA decision method based on the improved expert method is proposed to select the optimal allocation scheme, which can simultaneously take into account the maximum damage efficiency and the minimum ammunition consumption. After the two-stage attack, all five ships are seriously injured, which meets the operational requirements and verifies the rationality and effectiveness of the proposed method.

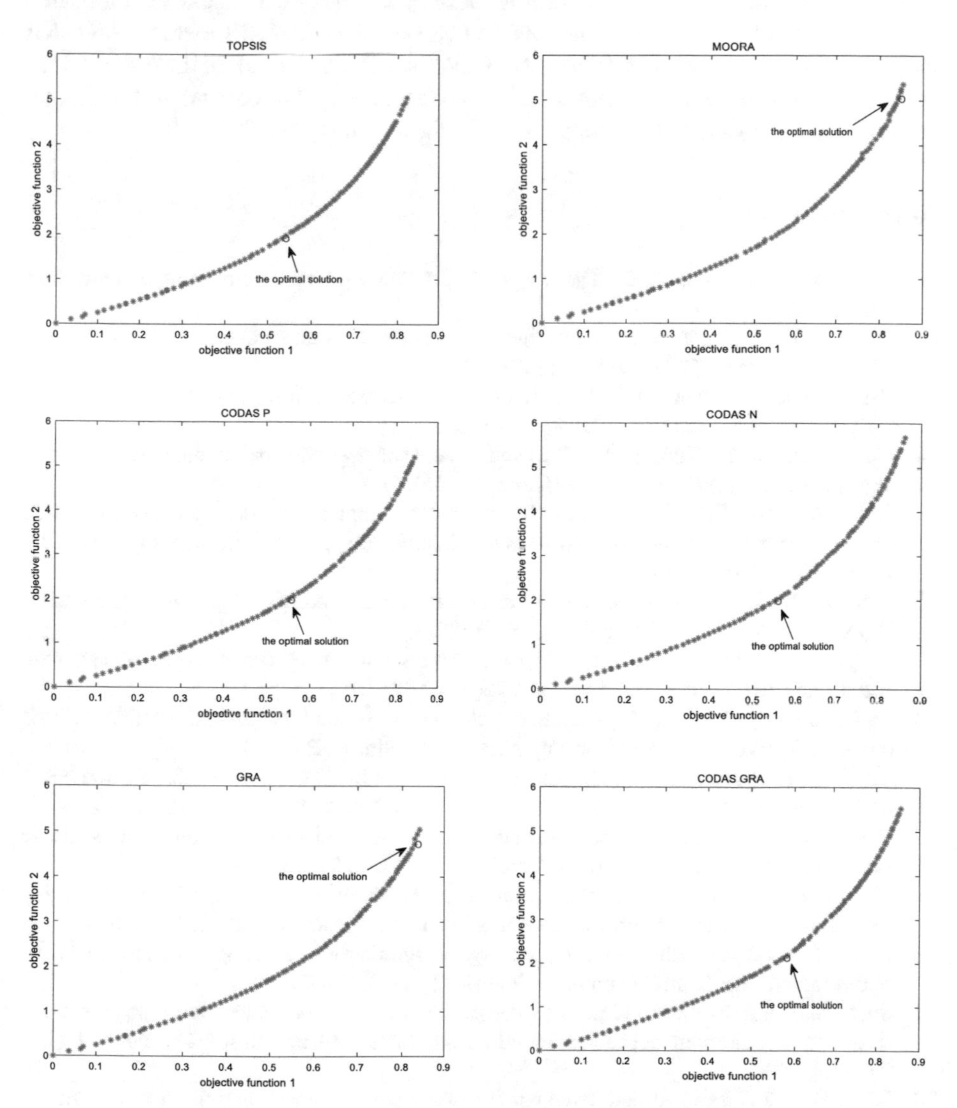

Fig. 4. The results of the WTA scheme selected by multi methods

6 Conclusion

In order to solve the projectile assignment problem of multi-type anti-ship missile combined attack ship formation, considering the constraints such as the total amount of ammunition, the number of fire channels, and the missile action distance, a multi-stage multi-objective optimization model with the maximum damage to the enemy and the minimum ammunition consumption was established, which was solved by using NSGA-II algorithm. A CODAS_GRA method based on the improved expert method is proposed to select the optimal weapon target allocation scheme. The effectiveness and rationality of the proposed method are verified by comparison with TOPSIS method, MOORA method, CODAS_P method, CODAS_N method and GRA method. In the next step, the application of deep reinforcement learning in weapon target allocation will be further studied, and the results will be compared with those in this paper.

References

1. Kline, A., Ahner, D., Hill, R.: The weapon-target assignment problem. Comput. Oper. Res. **105**, 226–236 (2019)
2. Gao, C., Kou, Y., Li, Y., et al.: Multi-objective weapon target assignment based on D-NSGA-III-A. IEEE Access **7**, 50240–50254 (2019)
3. Hu, X., Luo, P., Zhang, X., et al.: Improved ant colony optimization for weapon-target assignment. Math. Probl. Eng. **2018**, 1–14 (2018)
4. Ou, Q., He, X.Y., Guo, S.M.: Cooperative target assignment based on improved DDE algorithm. J. Command Control **5**(4), 282–287 (2019)
5. Zhu, J.W., Zhao, C.J., Li, X.P., et al.: Multi-target assignment and intelligent decision based on reinforcement learning. Acta Armamentarii. https://kns.cnki.net/kcms/detail/11.2176.TJ.20210512.1731.002.html
6. Yan, D., Su, H., Zhu, J.: Research on fire distribution method of anti-ship missile based on DQN. Navig. Position. Timing **6**(05), 18–24 (2019)
7. Ding, Z.L., Liu, G.L., Xie, Y., et al.: Dynamic targets assignment with reinforcement learning and neural network. Electron. Des. Eng. **28**(13), 7 (2020)
8. Li, K.W., Zhang, T., Wang, R., et al.: Research reviews of combinatorial optimization methods based on deep reinforcement learning. Acta Autom. Sin. 1–22 (2020)
9. Yu, B.W., Lyu, M.: Method for dynamic weapon coordinative firepower distribution based on improved NSGA-III algorithm. Fire Control Command Control **46**(8), 71–77 (2021)
10. Wang, C., Nan, Y., Xu, H.: A new elite genetic algorithm and its application in multi-missile interception assignment strategy. Aerosp. Control **39**(04), 59–66 (2021)
11. Li, J., Chen, J., Xin, B., et al.: Efficient multi-objective evolutionary algorithms for solving the multi-stage weapon target assignment problem: a comparison study. IEEE (2017)
12. Chen, M., Zhou, F.X.: Shipborne weapon target assignment based on improved particle swarm optimization. Fire Control Command Control **43**(11), 72–76 (2018)
13. Su, D.W., Wang, Y., Zhou, C.M.: Improved particle swarm optimization algorithm for solving weapon-target assignment problem based on intuitionistic fuzzy entropy. Comput. Sci. (12), 255–259 (2016)
14. Li, Y., Kou, Y., Li, Z.: An improved nondominated sorting genetic algorithm III method for solving multiobjective weapon-target assignment part I: the value of fighter combat. Int. J. Aerosp. Eng. **2018**, 1–23 (2018)

15. Wang, Z., Rangaiah, G.P.: Application and analysis of methods for selecting an optimal solution from the pareto-optimal front obtained by multiobjective optimization. Ind. Eng. Chem. Res. **56**(2), 560–574 (2017)
16. Wang, Z., Parhi, S.S., Rangaiah, G.P., et al.: Analysis of weighting and selection methods for pareto-optimal solutions of multiobjective optimization in chemical engineering applications. Ind. Eng. Chem. Res. **59**(33), 14850–14867 (2020)
17. Long, T., Liu, Z.Y., Shi, R.H., et al.: Neural network based air defense weapon target intelligent assignment method. Air Space Def. **4**(01), 1–7 (2021)
18. Xie, J.F., Yang, Q.M., Dai, S.L., et al.: Air combat maneuver decision based on reinforcement genetic algorithm. J. Northwest. Polytech. Univ. **38**(06), 1330–1338 (2020)
19. Feng, S., Zheng, B., Chen, W., et al.: RNSGA-II algorithm supporting reinforcement learning and its application in UAV path planning. Comput. Eng. Appl. **56**(03), 246–251 (2020)
20. Wei, H.X., Zhang, H.G.: Multi-target countermeasure method of a voiding attack based on deep neural network and ant colony algorithm. Comput. Appl. Softw. **37**(11), 7 (2020)
21. Chang, T., Kong, D., Hao, N., et al.: Solving the dynamic weapon target assignment problem by an improved artificial bee colony algorithm with heuristic factor initialization. Appl. Soft Comput. **70**, 845–863 (2018)
22. Deb, K., Pratap, A., Agarwal, S., et al.: A fast and elitist multiobjective genetic algorithm: NSGA-II. IEEE Trans. Evol. Comput. **6**(2), 182–197 (2002)
23. Keshavarz Ghorabaee, M., Zavadskas, E.K., Turskis, Z., et al.: A new combinative distance-based assessment (CODAS) method for multi-criteria decision-making. Econ. Comput. Econ. Cybern. Stud. Res. **50**(3) (2016)
24. Gao, F., Lin, H.H., Deng, H.W.: On the construction and application of the ground-water environment health assessment model based on the game theory-fuzzy matter element. J. Saf. Environ. **17**(4), 5 (2017)
25. Luukka, P., Collan, M.: Histogram ranking with generalised similarity-based TOPSIS applied to patent ranking. Int. J. Oper. Res. **25**(4), 437–448 (2016)

A Physical Ergonomics Study on Adaptation and Discomfort of Student's E-Learning in the Philippines During the COVID-19 Pandemic

Ryan M. Paradina and Yogi Tri Prasetyo(✉)

School of Industrial Engineering and Engineering Management, Mapúa University, 658 Muralla St., Intramuros, 1002 Manila, Philippines

ytprasetyo@mapua.edu.ph

Abstract. The e-learning or online learning is an education type that uses Internet. It is also common referred to as distance learning on which you will learn across distance and not in face to face classroom discussion. Because online learning involves various processes, the tasks of the students are faced with numerous body parts discomfort. The goal of this study is to identify and evaluate different discomforts and other potential risks in relation to student tasks using the knowledge on physical ergonomics. From this, the researcher was able to develop a recommendation to minimize these risks and discomfort. The current study applied descriptive analysis to analyze the impact among the students who are taking online classes at home and the number of hours they spent, and risk factors considered in this study and musculoskeletal disorder experienced by the students. Thirty students with combination of college, high school and senior high school were subjected to answer an online survey describing their position when taking online classes at home along with their own sample picture and Cornell Musculoskeletal Disorder Questionnaire (CMDQ). Key findings from the study revealed that students were exposed to discomfort particularly on Lower back, Upper Back and Hips/Buttocks. This is because of sitting between 7–8 h and 9–10 h a day when doing their online classes. To evaluate whole body postural discomfort, the researcher used the ergonomic assessment tool, Rapid Entire Body Assessment (REBA). Clearly, the researcher was able to identify risk related to health factor. From the result of the descriptive statistical analysis, factors like age and educational level affect the students to perform their task in having musculoskeletal disorder (MSD) or discomfort. Based from these conditions, the researcher provided a risk mitigation plan based from the results of the analysis to minimize the MSD or discomfort among the students.

Keywords: Discomfort · Musculoskeletal disorder · Online learning · Physical ergonomics

1 Introduction

In today's modern world and technology, the online education is now possible from every student wherein this will require a desktop computer or laptop and have access to the

© The Author(s), under exclusive license to Springer Nature Switzerland AG 2023
L. C. Tang and H. Wang (Eds.): BDET 2022, LNDECT 150, pp. 192–200, 2023.
https://doi.org/10.1007/978-3-031-17548-0_17

internet. Through this manner, it has changed the way of teaching among the instructors on how they will use presentation materials to their students [1]. In the Philippines, during the pandemic COVID-19 times the government or the Department of Education came up with a strategy that will infiltrate and adopt online learning into schools and universities. As a result of this situation, the increase of online study at home takes place wherein physical ergonomics is observed. Due this pandemic situation, students are tasked to stay at home and do the online learning. However, it also increases a physical ergonomic discomfort since student will tend to sit only for couple of hours a day or per online class session. Increasing the task load will have significant effect on the performance of the students [2].

One of the problems is the exposure of the students mostly to sitting for longer hours at home while studying online. The activities in online learning involves health risks to the students from the point they started sitting and the impact of doing this to their body [3]. As a student, it is not good when you feel discomfort while a certain subject is on-going and learning is not focus. This study will show also the students demographics success in online learning. It will also focus on the relationship of the student using his computer and the effect of his performance. When we understand these relationships, it will be helpful in the future on a successful online learning environment. There are several studies and researches have been developed and published regarding the possible harm and risk of steady sitting throughout the online class [4–7]. Most researches about physical ergonomics on online learning are focused on the task and completing the activities of the students but not giving attention to the proper posture, the way they sit and the equipment used.

According to a research published International Review of Research in Open and Distributed Learning, the online and digital learning were predicted to go up as we move towards the modern world. It discusses the issues and challenges that open and distance e-learning (ODeL) poses for the Philippines' open university from the point of view of the institution's leading ODeL practitioners [8].

A medical study and research particularly on musculoskeletal disorders pointed out related to this paper, major findings are pain in the neck and lower back areas followed by thigh pain. During this time of preparation for the competitive exams, most students have altered and irregular eating habits, sit in improper body posture for long hours, neglect regular physical exercise, have the stress of staying away from home or studying alone for long hours, etc. All this makes the students prone for musculoskeletal pains. While many students seek medical help, most neglect their symptoms or procrastinate until the exams [9].

The purpose of this study is to obtain the relationship between physical ergonomic discomfort, the task and workload, and the impact of physiological response to student performance while using their computers as a tool for distance learning to read course material at home or at their own space.

When this paper is released, it will contribute to the physical ergonomics study for a specific related topic on digital learning at home. This paper will benefit to teachers and related academic institutions by showing academic strategic plan and mitigation plan on how to apply an effective physical ergonomics for online education or distance learning. Furthermore, this study will support in understanding the limitations on physical ergonomics and expectations for students that will help guide on what will be the presentation materials, course program and design, and attributes of computer or laptops, tables and chairs that are suitable for online learning environments.

2 Methodology

The researcher conducted a study among three different student levels: high school, senior high school and college wherein these students are common to be engaged in online digital learning. A total of 30 students (respondents) with breakdown of 20 college, 8 senior high school and 2 high school (Table 1). An online survey is conducted and sent to the students in order to obtain demographic data, pictures to provide RULA and REBA scores, and CMDQ to identify where are the location of discomfort and other common types of musculoskeletal disorders experienced by the students and assess whole body postural MSD then measure what are the risks associated while performing the task which sitting throughout their online class (Fig. 1).

Table 1. Demographics.

Demographic	Description	Count	Percentage
Students	Total	30	100%
Gender	Female	24	80%
	Male	6	20%
Age	24 years old	6	20%
	23 years old	6	20%
	22 years old	4	13%
	21 years old	2	7%
	20 years old	2	7%
	19 years old	4	13%
	18 years old	1	3%
	17 years old	3	10%
	16 years old	1	3%
	13 years old	1	3%
Educational Level	College	20	67%
	High School	2	7%
	Senior High School	8	27%
Computer	Desktop	8	27%
	Laptop	22	73%

(*continued*)

Table 1. (*continued*)

Demographic	Description	Count	Percentage
Study Hours	1–2 h a day	1	3%
	3–4 h a day	2	7%
	5–6 h a day	2	7%
	7–8 h a day	9	30%
	9–10 h a day	9	30%
	10 h and above	7	23%

Fig. 1. One subject being measured.

2.1 Selecting a Template

The statistical analysis on this study is descriptive measures in order to provide summary of factors based from the data collection. These collected data from survey were arranged in a simple manner on which the data can be easily understand and illustrated through the use of frequency count and percentage distribution. In addition, the factors considered in the study in processing the data is the standard statistical methods on which it is used for calculating the mean and the standard error (±SE).

3 Results and Discussion

Table 2 below shows overall result summary using CMDQ from the 30 students and ranked from highest to lowest risk following the scoring guidelines [10].

Table 2. CMDQ summary table.

Body parts	N	Discomfort	Interfe-rence	Discomfort score	%
Lower Back	82	48	43	169248	15.96
Hip/Buttocks	62.5	42	41	107625	10.15
Upper Back	68.5	41	38	106723	10.07
Neck	62	40	39	96720	9.12
Wrist (R)	49.5	43	41	87269	8.23
Shoulder (R)	60	36	37	79920	7.54
Shoulder (L)	48	34	33	53856	5.08
Lower Leg (R)	42	36	33	49896	4.71
Lower Leg (L)	42	36	33	49896	4.71
Upper Arm (R)	36	36	35	45360	4.28
Forearm (R)	29	36	34	35496	3.35
Knee (R)	30.5	35	33	35228	3.32
Knee (L)	29	35	32	32480	3.06
Thigh (R)	24	32	32	24576	2.32
Thigh (L)	24	32	32	24576	2.32
Wrist (L)	20.5	36	33	24354	2.30
Upper Arm (L)	22.5	33	31	23018	2.17
Forearm (L)	14	32	31	13888	1.31

According to the total discomfort score of Table 2, it was concluded that students felt discomfort mostly on lower back (15.96%), hip/buttocks (10.15%) and upper back (10.07%) while it was less pronounced in the forearm (1.31%), left upper arm (2.17%) and left wrist (2.30%).

More specifically, the results indicated that 14 students (46.67%) sensed discomfort in lower back 1–2 times per week and had an effect on their ability to study. Eleven (36.67%) students assessed hip/buttocks discomfort 1–2 times per week and had an effect on their ability to study. Twelve students had discomfort on upper back 1–2 times per week and had an effect on their ability to study.

From the results of the data, majority of the discomfort encountered by the students obtained from sitting at home is their lower back. This is because of their proper posture when sitting, the chair used is not adequate to use for longer hours and the reach of their laptop or computers that their body was stretched. A Rapid Entire Body Assessment (REBA) was conducted in order to validate the cause of discomfort or musculoskeletal disorders of students in sitting and doing online study [11, 12]. Table 3 below shows the highest REBA score.

The results on the table above showed that this type of posture for sitting while doing the online study task has a very high risk for students that will develop MSD specially to

Table 3. Highest REBA score.

A. Neck, Trunk and Leg Analysis		B. Arms and Wrist Analysis	
Factor	Score	Factor	Score
Neck Score	2	Upper Arm Score	3
Trunk Score	2	Lower Arm Score	2
Leg Score	2	Wrist Score	2
Posture Score A	4	Posture Score B	5
Force/Load Score	0	Coupling Score	1
Score A	4	Score B	6
Table C Score 6			
Activity Score 2			
REBA Score 8			

lower back, hip/buttocks and upper back. If maintained at this position for a longer period of time, it proves why majority of the students experienced MSDs in these mentioned parts of the body. Using CMDQ, the scores of the students are also at high. This sitting posture is very common to the students since they only utilize what they have and is available at their home.

3.1 Risk Mitigation for Sitting Task (While Doing Online Class)

Based on the CMDQ and REBA scores, there is need to recommend and implement mitigation on the identified risk factors on lower back, hip/buttocks and upper back. It was determined in the study that students obtained discomfort from these parts of the body due to continuous sitting and pressure on their online classes and lack of other physical activities. It can be suggested to do physical stretching and warm up and prior to online class and repeat again when a discomfort is felt.

- For the lower back and upper back discomforts, take micro breaks from sustained posture after 30–40 min and stretch prior to the sitting task which is helpful to the students and change the position whenever felt discomfort. This will help also to at least eliminate the body point pressure.
- Hip/buttocks must perform brief stretching and warm up, stand at least and walk 5–10 min inside their house.

3.2 Recommendation on Ergonomic Workstation for Studying

Physical ergonomics doesn't have to be expensive and should use the available equipment available at home.

Table or desk set up – If you already have an available table or desk at home, you can gage, measure the distance and enough space between your knees and under the table.

Make sure that your arms and hands are at rest and at level as you use your keyboard and not reaching too far. This obviously will result an automatic discomfort between these body parts when prolong. Other designs of the tables or desks already integrated with keyboard that will allow you to change the height depending on your size and reach. This desk type has beneficial to the user wherein the keyboard can be move closer or farther within the reach of the user. It also makes you feel comfortable when your feet is flat on the ground and could prevent neck and shoulder discomforts.

Chair – The available chair at home can be utilized as long as you sit properly and comfortably. You can add a soft pillow at your back and neck. There are ergonomic chairs that is available in the market that you can buy if you have budget. It can be adjustable and well fitted to the user. However, it is still recommended that you need to be resourceful of the chair available at home. This should satisfy the upper and lower backs of the body that will prevent discomfort.

Monitor to avoid eye strain – For the monitor, you can adjust the brightness level until you are satisfied and your eyes felt warm. Changing the colors of your screen may not be necessary and not suitable as it may cause you eye strain discomfort. Just remain the normal monitor display settings.

Keyboard designs – Normal keyboard designs that we can see together with our computer is common. There are two types: a flat and a split angled. Both is concerned on the ergonomic position of the depending on the user. Keyboards should be at level on your arm length and by reaching it in order to perform comfortably. When the position of the keyboard is not properly placed, your shoulder and back will cause you discomfort.

Mouse – The purpose of mouse is to hover and click an item on your computer during the online class. A common mouse is placed on top of the table beside your keyboard. This means that the level and reach of the mouse should be the same as the keyboard level. Some of the ergonomic mouse design are expensive. Whenever you feel discomfort on your hands or wrist, take a rest first and do some hand stretching. The mouse should be enough size on your hand. This will prevent wrist and grip discomforts.

3.3 Limitations of This Study

One of the limitations of this study is the number of students as respondents. The study could further analyze with more respondents targeting minimum of 100 students and maximum of 300 students as only 30 students were used in this study. The use of pictures in order to evaluate the REBA for proper scoring and can extend up to utilizing RULA as well. The time limitation to conduct this study is only within 8 weeks.

4 Conclusion

Ergonomics is an important part of [13–27] even during the COVID-19 pandemic. In the end of this study, it was found that students doing online study at home are susceptible to body and health risk factors. Because of this, it leads to physical discomforts associated with MSD of students. It was identified that students are not aware of their sitting posture and has no enough knowledge about physical ergonomics while performing the tasks and doing their online class. These were caused by having no appropriate tools available at their homes that they can use or whatever available equipment found at home for online study set up activities like desk table, proper location and set up for their desktop computer or laptop, proper handling of the mouse, and most especially the available chair as it entails the most body discomforts mentioned in this study (lower back, hip/buttocks and upper back) as there is no back support. The students also experienced discomforts due to longer hours of sitting caused by the need of the online study (7–8 h and 9–10 h). More so, it was shown on the result of the analysis that health and body risk related factors contributed to the discomfort and musculoskeletal disorders of students in sitting task. Clearly, the elements that were identified to have high contribution to the discomforts of students are demographic profile, CMDQ and REBA. Due to these conditions, the researcher was able to come up with the risk mitigation that includes recommendation to ease MSD on lower back, hip-buttocks and upper back, and recommendation on how to be resourceful on your available equipment at home in order to perform sitting task while doing the online classes. For the future research related to this study, the next researcher can acquire and make study of the relationship between musculoskeletal discomfort versus the quality of the performance and habits of the students doing the online tasks.

Acknowledgment. The researcher would like to acknowledge all students who participated on this study and who gave lots of their time and effort on providing the information needed.

References

1. Gilbert, B.: Online Learning Revealing the Benefits and Challenges. St. John Fisher College, Education Masters (2015)
2. Jones, R.: Physical Ergonomic and Mental Workload Factors of Mobile Learning Affecting Performance of Adult Distance Learners: Student Perspective. University of Central Florida, Electronic Theses and Dissertations, 2004–2019 (2009)
3. Fraser, J., Holt, P., Mackintosh, J.: Online learning environments: a health promotion approach to ergonomics. In: Association for the Advancement of Computing in Education (AACE), Chesapeake, VA, Athabasca University, Canada. Texas Publisher, San Antonio (2000)
4. Lakarnchua, O., Balme, S., Matthews, A.: Insights from the implementation of a flipped classroom approach with the use of a commercial learning management system. Turk. Online J. Distance Educ. **21**(3), 63–76 (2020)
5. Okulova, L.: An ergonomic approach to higher education of psychology and pedagogy students. **41**(02), 13 (2020). ISSN 0798 1015
6. Harrati, N., Bouchrika, I., Tari, A., Ladjailia, A.: Exploring user satisfaction for e-learning systems via usage-based metrics and system usability scale analysis. Department of Computer Science, University of Bejaia, 6000, Algeria, Faculty of Technology and Science, University of Souk Ahras, 41000, Algeria (2016)
7. Dela Pena-Bandalaria, M.: E-learning in the Philippines: trends, directions, and challenges. Int. J. E-Learn. **8**(4) (2009). ISSN 1537-2456
8. Arinto, P.: Issues and challenges in open and distance e-learning: perspectives from the Philippines. University of the Philippines, vol. 17, no. 2 (2016)
9. Santoshi, J., Jain, S., Popalwar, H., Pakhare, A.: Musculoskeletal disorders and associated risk factors in coaching students: a cross-sectional study (2019)
10. Hedge, A., Morimoto, S., McCrobie, D.: Effects of keyboard tray geometry on upper body posture and comfort. Ergonomics **42**(10), 1333–1349 (1999)

11. Hignette, S., McAtamney, L.: Rapid entire body assessment: REBA. Appl. Ergon. **31**, 201–205 (2000)
12. Cakit, E.: Ergonomic risk assessment using Cornell musculoskeletal discomfort questionnaire in a grocery store. Age **3**(6) (2019). ISSN 2577-2953
13. Cooper, K.N., Sommerich, C.M., Campbell-Kyureghyan, N.H.: Technology in the classroom: College Students' computer usage and ergonomic risk factors (2009)
14. Jansen, K., et al.: Musculoskeletal discomfort in production assembly workers. Acta Kinesiologiae Universitatis Tartuensis **18**, 102–110 (2012)
15. Erdinca, O., Hotb, K., Ozkayac, M.: Turkish version of the Cornell Musculoskeletal Discomfort Questionnaire: cross-cultural adaptation and validation. Work **39**, 251–260 (2011). https://doi.org/10.3233/WOR-2011-1173
16. Gumasing, M.J., et al.: Ergonomic design of apron bus with consideration for passengers with mobility constraints. Safety **8**, 33 (2022)
17. Gumasing, M.J., Prasetyo, Y.T., Ong, A.K., Persada, S.F., Nadlifatin, R.: Analyzing the service quality of E-trike operations: a new sustainable transportation infrastructure in Metro Manila Philippines. Infrastructures **7**, 69 (2022)
18. Almutairi, N.A., Almarwani, M.M.: Knowledge and use of the International Classification of functioning, Disability and Health (ICF) and ICF core sets for musculoskeletal conditions among Saudi physical therapists. Musculoskeletal Sci. Pract. **60**, 102573 (2022)
19. Yazdanirad, S., Pourtaghi, G., Raei, M., Ghasemi, M.: Developing and validating the personal risk assessment of Musculoskeletal Disorders (PRAMUD) tool among workers of a steel foundry. Int. J. Ind. Ergon. **88**, 103276 (2022)
20. Wang, S., Wang, Y.: Musculoskeletal model for assessing firefighters' internal forces and occupational musculoskeletal disorders during self-contained breathing apparatus carriage. Saf. Health Work (2022). https://doi.org/10.1016/j.shaw.2022.03.009
21. Kox, J., et al.: What sociodemographic and work characteristics are associated with musculoskeletal complaints in nursing students? A cross-sectional analysis of repeated measurements. Appl. Ergon. **101**, 103719 (2022)
22. Oestergaard, A.S., Smidt, T.F., Søgaard, K., Sandal, L.F.: Musculoskeletal disorders and perceived physical work demands among offshore wind industry technicians across different turbine sizes: a cross-sectional study. Int. J. Ind. Ergon. **88**, 103278 (2022)
23. Albanesi, B., Piredda, M., Bravi, M., et al.: Interventions to prevent and reduce work-related musculoskeletal injuries and pain among healthcare professionals. A comprehensive systematic review of the literature. J. Saf. Res. (2022). https://doi.org/10.1016/j.jsr.2022.05.004
24. Muhamad Hasani, M.H., Hoe Chee Wai Abdullah, V., Aghamohammadi, N., Chinna, K.: The role of active ergonomic training intervention on upper limb musculoskeletal pain and discomfort: a cluster randomized controlled trial. Int. J. Ind. Ergon. **88**, 103275 (2022)
25. Mohamaddan, S., et al.: Investigation of oil palm harvesting tools design and technique on work-related musculoskeletal disorders of the upper body. Int. J. Ind. Ergon. **86**, 103226 (2021)
26. Miller, A.T., et al.: Procedural and anthropometric factors associated with musculoskeletal injuries among gastroenterology endoscopists. Appl. Ergon. **104**, 103805 (2022)
27. Fouladi-Dehaghi, B., Tajik, R., Ibrahimi-Ghavamabadi, L., Sajedifar, J., Teimori-Boghsani, G., Attar, M.: Physical risks of work-related musculoskeletal complaints among quarry workers in east of Iran. Int. J. Ind. Ergon. **82**, 103107 (2021)

Author Index

© The Editor(s) (if applicable) and The Author(s), under exclusive license
to Springer Nature Switzerland AG 2023

L. C. Tang and H. Wang (Eds.): BDET 2022, LNDECT 150, pp. 201–202, 2023.
https://doi.org/10.1007/978-3-031-17548-0